Borders: A Very Short Introduction

VERY SHORT INTRODUCTIONS are for anyone wanting a stimulating and accessible way into a new subject. They are written by experts, and have been translated into more than 45 different languages.

The series began in 1995, and now covers a wide variety of topics in every discipline. The VSI library currently contains over 750 volumes—a Very Short Introduction to everything from Psychology and Philosophy of Science to American History and Relativity—and continues to grow in every subject area.

Very Short Introductions available now:

ABOLITIONISM Richard S. Newman
THE ABRAHAMIC RELIGIONS
    Charles L. Cohen
ACCOUNTING Christopher Nobes
ADDICTION Keith Humphreys
ADOLESCENCE Peter K. Smith
THEODOR W. ADORNO
    Andrew Bowie
ADVERTISING Winston Fletcher
AERIAL WARFARE Frank Ledwidge
AESTHETICS Bence Nanay
AFRICAN AMERICAN HISTORY
    Jonathan Scott Holloway
AFRICAN AMERICAN RELIGION
    Eddie S. Glaude Jr.
AFRICAN HISTORY John Parker and
    Richard Rathbone
AFRICAN POLITICS Ian Taylor
AFRICAN RELIGIONS
    Jacob K. Olupona
AGEING Nancy A. Pachana
AGNOSTICISM Robin Le Poidevin
AGRICULTURE Paul Brassley and
    Richard Soffe
ALEXANDER THE GREAT
    Hugh Bowden
ALGEBRA Peter M. Higgins
AMERICAN BUSINESS HISTORY
    Walter A. Friedman
AMERICAN CULTURAL HISTORY
    Eric Avila
AMERICAN FOREIGN RELATIONS
    Andrew Preston
AMERICAN HISTORY Paul S. Boyer

AMERICAN IMMIGRATION
    David A. Gerber
AMERICAN INTELLECTUAL HISTORY
    Jennifer Ratner-Rosenhagen
THE AMERICAN JUDICIAL SYSTEM
    Charles L. Zelden
AMERICAN LEGAL HISTORY
    G. Edward White
AMERICAN MILITARY HISTORY
    Joseph T. Glatthaar
AMERICAN NAVAL HISTORY
    Craig L. Symonds
AMERICAN POETRY David Caplan
AMERICAN POLITICAL HISTORY
    Donald Critchlow
AMERICAN POLITICAL PARTIES
    AND ELECTIONS L. Sandy Maisel
AMERICAN POLITICS
    Richard M. Valelly
THE AMERICAN PRESIDENCY
    Charles O. Jones
THE AMERICAN REVOLUTION
    Robert J. Allison
AMERICAN SLAVERY
    Heather Andrea Williams
THE AMERICAN SOUTH
    Charles Reagan Wilson
THE AMERICAN WEST Stephen Aron
AMERICAN WOMEN'S HISTORY
    Susan Ware
AMPHIBIANS T. S. Kemp
ANAESTHESIA Aidan O'Donnell
ANALYTIC PHILOSOPHY
    Michael Beaney

THE WORLD TRADE
    ORGANIZATION
    Amrita Narlikar
WORLD WAR II Gerhard L. Weinberg

WRITING AND SCRIPT
    Andrew Robinson
ZIONISM Michael Stanislawski
ÉMILE ZOLA Brian Nelson

Available soon:

DOSTOEVSKY
    Deborah Martinsen
THUCYDIDES Jennifer T. Roberts

BRITISH ARCHITECTURE
    Dana Arnold
SUSTAINABILITY Saleem H. Ali

For more information visit our website

www.oup.com/vsi/

Alexander C. Diener and Joshua Hagen

BORDERS

A Very Short Introduction

SECOND EDITION

OXFORD
UNIVERSITY PRESS

Oxford University Press is a department of the University of Oxford.
It furthers the University's objective of excellence in research, scholarship,
and education by publishing worldwide. Oxford is a registered trade mark of
Oxford University Press in the UK and in certain other countries.

Published in the United States of America by Oxford University Press
198 Madison Avenue, New York, NY 10016, United States of America.

Library of Congress Cataloging-in-Publication Data

Names: Diener, Alexander C., author. | Hagen, Joshua, 1974- author.
Title: Borders : a very short introduction / Alexander C. Diener and Joshua Hagen.
Description: Second edition. | New York : Oxford University Press, 2024. |
Includes bibliographical references and index.
Identifiers: LCCN 2023039557 (print) | LCCN 2023039558 (ebook) |
ISBN 9780197549605 (paperback) | ISBN 9780197549629 (epub)
Subjects: LCSH: Boundaries. | Borderlands. | Boundary disputes. |
Human territoriality—Political aspects. | Human geography. |
Political anthropology. | International relations.
Classification: LCC JC323 .D54 2024 (print) | LCC JC323 (ebook) |
DDC 320.1/2—dc23/eng/20230828
LC record available at https://lccn.loc.gov/2023039557
LC ebook record available at https://lccn.loc.gov/2023039558

Integrated Books International, United States of America

Contents

List of illustrations

Acknowledgments

We would like to thank everyone at Oxford University Press, especially Nancy Toff, for their work to bring this second edition to press. We also thank our families Bethany and Maximus Diener, and Sabina and Oliver Hagen, for their continuing support.

Alexander would also like to acknowledge that this project was completed while he was a visiting fellow at the University of Connecticut's Humanities Institute (2023–2024).

Chapter 1
A very bordered world

The world is filled with borders. Some borders are highly visible and figure prominently in discussions of national security, migration, trade, natural resources, and international relations. Other borders may be less visible but are equally impactful in domestic affairs and shape elections, land use, property rights, and senses of belonging, among many other issues. As a result, borders are taken as "natural" and unquestioned realities of daily life. Rather than being natural phenomena, this Very Short Introduction demonstrates that borders result from human action and thought, specifically our propensity to organize space and place and thereby create geographic frameworks for understanding our place in the world.

Humans have organized space for millennia—perhaps from our beginnings as a species—but a variety of territorial strategies have been pursued to address a wide range of political, socioeconomic, environmental, and technological circumstances. That variety gradually coalesced around certain global norms, standards, and expectations for borders over the past several centuries. This convergence began first in medieval Europe when a convoluted hodgepodge of feudal, municipal, and ecclesiastical territorial hierarchies gradually evolved into nation-states exercising sovereignty over territories with well-defined borders. States simultaneously sought to establish greater domestic control

1. Workers erect a new fence along the U.S.-Mexico border near El Paso, Texas, ca. 2011.

through expanding bureaucracies and the creation of subordinate jurisdictions, such as provinces, counties, townships, municipalities, and other local authorities. At times, these external and internal borders worked to sharpen ethnic, cultural, or religious differences, while in other instances borders served to homogenize societies. This new model of organizing political space based on territorial sovereignty was subsequently imposed on the rest of the world through European colonial conquest, most prominently during the eighteenth and nineteenth centuries. By the twentieth century, territorial perspectives and assumptions regarding politics, culture, and socioeconomics had come to pervade daily life.

Most people cross hundreds of borders every day, but borders are most obvious when they block movement, whether through something as simple as "No Trespassing" or "Private Property" signs or something more substantial like the walls, fences, and checkpoints that commonly mark international borders. From international to local scales, borders permeate daily life by creating and formally demarcating different legal, political, and economic spaces, such as electoral districts, court jurisdictions, and gated communities. Less obvious but no less important, borders also shape informal social and cultural spaces, such as ethnic neighborhoods, gendered spaces, and gang territories, among others.

Despite that ubiquity, borders are *not* natural phenomena. Humans may be predisposed to organize space and create places, but how we structure territory, and to what end, has varied quite radically over time. Those studying borders and the processes of bordering come from a wide range of scholarly perspectives and constitute a growing and dynamic field. We hope that this book's survey of border history and border research promotes greater awareness, understanding, and further study.

Territory, sovereignty, and borders

We most commonly associate borders with lines on a map or a collection of walls, fences, signs, and checkpoints separating two independent countries, but their most basic function is to establish and maintain different places and spaces. In other words, borders separate the meaning and function of one geographic area from another. The world is full of various types of boundaries, but the word borders is normally associated with the idea of territory, or a geographic area intended to regulate the movement of people and things while also conveying certain behavioral expectations. That propensity to create territories manifests in various modes of territoriality through which people create, communicate, and control geographical spaces, either individually or collectively. Territoriality finds expression in varied modalities ranging from the placement of permanent markers and the precise demarcation of cartographic lines to the performance of periodic ceremonies and the vague designation of frontier zones. Therefore, territoriality and bordering strategies are highly contingent, adaptable, and mobile, rather than constant, consistent, and fixed, and have varied significantly over time and across space.

The root causes of territoriality remain hotly debated by scholars. Drawing from sociobiology, some scholars have favored "primordialist" approaches that regard territoriality as rooted in an a priori instinct. In this view, social groups instinctively seek territorial control to secure resources necessary for survival. This approach suggests that humans compete in perpetual "survival of the fittest" contests as groups seek to control territory, secure resources, and regulate access of rival groups. Although animals exhibit territoriality, for example by marking out hunting ranges with scent, attempts to liken human territoriality to a base primordial instinct unduly relegate a very complex process to a natural reflex.

Human place-making and territoriality differ from that of animals in two distinct ways. First, territorial control is not, nor has it ever been, the sole means by which humans exercise social control. Countless forms of de-territorialized "authority"—or the legitimate exercise of power—have existed throughout history and continue today. Contemporary examples include various religious and social movements, as well as nongovernmental organizations focused on environmentalism, human rights, and feminism, that champion their respective causes as universal and therefore possessing authority across political and socioeconomic categories. The global influence of certain businesses, such as social media or information technology giants, could also be considered a form of de-territorialized authority since their platforms clearly transcend territorial boundaries.

The second difference between human and animal territoriality is the flexibility and malleability of human territoriality. Unlike animals, human spatial thinking has encompassed a wide variety of strategies, assumptions, practices, and perspectives over time.

Frontiers—or ill-defined regions with limited rule of law and unclear sovereignty—were once common and accepted but are increasingly rare and generally regarded as problems today. Also, some cultures did not develop notions of property and land ownership until confronted with land claims from outside groups. The recurring clashes throughout history between pastoral nomads and sedentary farmers vividly illustrate the incongruity between communal or public land-use and private land-rights. Both groups engaged in territoriality, but they conceptualized and practiced territoriality in very different ways.

The changing nature of human spatiality—or human conceptualizations and interactions with space—leads to alternative explanations for territoriality. In contrast to the primordialist approach, many scholars favor a constructivist approach that regards unequal power relations as the primary catalysts for creating categories of "us" and "them," "insiders" and "outsiders," and "in place" and "out of place." Territoriality, therefore, serves as a mechanism of social control, driving an ongoing negotiation over rightful possession of land. By demarcating and defending territory, groups control specific spaces as a means to regulate movement, access to resources, and behavioral expectations. These territorial practices can be altered to fit a variety of political, socioeconomic, and cultural agendas, thus lending flexibility, malleability, and ambiguity to identities and their associated places. Regardless of its origin, territoriality has become increasingly institutionalized and globalized over the past several centuries, so much so that the blatantly social and political processes of bordering appear natural.

Borders are manifestations of territoriality that assign people, things, and activities to particular places and in doing so constitute a spatial strategy to regulate mobility and norms associated with those people, things, and activities. These inherently social and political processes contribute to notions of ownership—or the rightful and permanent possession of land—whether by an individual, family, business, government, or other social groupings. Over recent centuries, the development of dramatic disparities in power within and between social groups has given rise to the related concepts of sovereignty and jurisdiction. Though rather recent inventions, these concepts were central to defining the territorial limits of government power and establishing borders as organizing principles for the modern state system.

Sovereignty is defined as the exercise of supreme authority and control over a distinct territory and its corresponding population and resources. Jurisdiction refers to the subordinate authority of a particular person, group, or institution within a sovereign authority. Both are complex ideas that help structure the spatiality of governance—state, ethnic territory, province, municipality, etc.—and the exercise of political power. Sovereignty and jurisdiction imply general recognition of authority over a bounded territory while also obscuring the often-violent processes that supplanted alternative political frameworks. Because identity is malleable and spatial patterns of interaction and exchange are fluid, absolute territorial sovereignty is impossible to realize in practice. Not even the famed Iron Curtain completely blocked the flow of people, goods, and ideas. As such, territorial sovereignty, in conjunction with national territoriality, serves as both cause and effect in many domestic and international controversies.

Today, perhaps more than ever, the rise of cross-border processes, patterns, and problems—often lumped together under the term "globalization"—challenges established notions of territorial sovereignty. By their very nature, environmental impacts resulting from pollution or climate change transcend territorial boundaries. One country's efforts to "go green" may be undermined by a neighbor's dirtier industrial practices that affect common air and water quality. In economics and business, transnational corporations increasingly benefit from common markets (for example, the European Union, COMESA, MERCOSUR,

CARICOM, ASEAN) and lower tariffs as they build supply chains and distribution networks that span the world. Global networks of information, innovation, investment, and education now link the East to the West and the North to the South so tightly that the gap between "haves" and "have-nots" is a conspicuous fact of life. Numerous supranational institutions and private groups work across borders to narrow these gaps, but they face daunting challenges. Dual, multiple, and contingent citizenship are also increasingly sought among the highly mobile members of various societies, ranging from wealthy cosmopolites jet-setting between global cities to migrants fleeing war, famine, environmental degradation, oppression, or poverty.

In the early twenty-first century, following the 9/11 terrorist attacks, border research tended to focus on terrorism and other criminal activities. Today, border research tackles a broader range of topics, including robust debates over migration (as a threat versus human right or economic asset), the effects of populism (the United Kingdom's withdrawal from the European Union, protectionist trade policies), and blatant land-grabs (Ukraine, South China Sea). Rather than moving toward a borderless world as some predicted in the 1990s, many states have instead strengthened border security to better control flows into and out of their territories.

Some of the most prominent examples include the United States constructing hundreds of miles of fencing along its border with Mexico; India fencing its 2,500-mile (4,000-km) border with Bangladesh and 1,800-mile (2,900-km) border with Pakistan; Pakistan building fences and laying minefields along sections of the Afghanistan border; and Iran walling its 430-mile (700-km) border with Pakistan. Israel constructed a 440-mile (708-km) security barrier around many Palestinian areas in the West Bank, while shorter fences have been built along Israel's borders with Gaza and Egypt. Beyond these highly publicized examples, numerous other countries—including Algeria, Argentina, Austria,

Azerbaijan, Botswana, Brunei, Bulgaria, China, Denmark, Dominican Republic, Egypt, Estonia, Finland, France, Greece, Hungary, Iraq, Jordan, Kazakhstan, Kenya, Kuwait, Kyrgyzstan, Latvia, Lithuania, Malaysia, Morocco, Myanmar, Nigeria, North Macedonia, Norway, Oman, Poland, Russia, Saudi Arabia, Serbia, Slovenia, South Africa, Spain, Thailand, Tunisia, Turkey, Turkmenistan, Ukraine, United Arab Emirates, United Kingdom, and Uzbekistan—have launched new border wall and fence construction projects since 2000. These recent projects add to older border barriers, such as between South Korea-North Korea, Cyprus-North Cyprus, and the so-called "berm" partitioning Western Sahara.

The field of border research

Philosophical debates regarding the nature of an ideal society have consistently explored the tension between openness and closedness. In Plato's *Laws* from around 360 BCE, Sparta is presented as a model polity, which emphasizes "security" in pursuit of the ultimate goals of virtue and happiness. "Opportunity" in the form of territorial expansion, maritime expeditions, and distant resource exploration was to be approached with caution, if not mistrust, even if it promised substantial profits. In contrast, Aristotle's *Politics* from around 350 BCE critiques Plato by calling for balance between security and opportunity. From this perspective, isolation was counterproductive since the polity required engagement and exchange with external partners to ensure its prosperity and very survival. This same basic debate has continued through the centuries—Thomas Hobbes (*Leviathan* in 1651 CE) and Adam Smith (*The Wealth of Nations*, 1776) to Thomas Friedman (*The World Is Flat*, 2005) and Harm de Blij (*The Power of Place*, 2008) and John Agnew (*Hidden Geopolitics*, 2022)—attesting that political systems are neither completely closed as idealized by rigid notions of territorial sovereignty nor completely open in some sort of "end of geography" resulting from flows of globalization.

Reflecting their centrality to social relations and the exercise of political power, borders have become hot research topics across several disciplines. Since the late 1960s, geographers, sociologists, anthropologists, economists, environmental psychologists, political scientists, legal scholars, and historians have challenged prior scholarship presuming that borders play rather passive roles in international and domestic affairs. Rather than simple pretexts for conflict or impediments to mobility, scholars began to consider borders as key geopolitical processes and mechanisms of social control worthy of deeper consideration across multiple scales. More recently, a growing number of politicians, business leaders, and the general public have questioned the assumed fixed relationship between power and territory inherent in the nation-state system, recognizing that very different modes of territorial organization existed around the world merely three centuries ago. The political order resulting from the transition of monarchical rule by hereditary nobility to democratic governments of elected representatives, along with urbanization, industrialization, and the diffusion of "modernity," is a relatively recent and spatially uneven phenomenon. Nevertheless, these events gave rise to new territorial assumptions and practices that had profound effects on the way people perceive themselves and their place in the world.

The replacement of vague frontiers with clearly defined borders reframed human identity and most social processes by ensnaring them in what political geographers commonly call the "territorial trap." This concept derives from three interrelated assumptions. The first is that states exercise exclusive and absolute power over their territories. In other words, states possess sovereignty. The second assumption regards domestic and international (internal and external) affairs as different realms of political and socioeconomic activity with fundamentally different standards of legality and morality. The third assumption views the political borders of the state as matching the boundaries of society and the economy. In other words, states serve as rigid containers that neatly partition global space into nation-state territories corresponding to distinct polities, societies, economies, and even environments.

Combined with the emotional appeal of nationalism, these assumptions reinforced a state-centric view of power and a corresponding partitioning of global space. As political, socioeconomic, and cultural practices became increasingly associated with the state during the nineteenth century, business, labor, politics, sports, military, education, and the arts came to be viewed through the prism of the territorial trap. The notion that those things were most efficient, practical, and logical when conceptualized and organized in relation to the state was "scientifically" supported through the new social sciences of economics, sociology, and political science. The rise of the modern academy was very much in service to the nation-state. As organized religion receded as the primary educator of the populace, the framing and shaping of identity to fit the new religion of the nation—that is nationalism—fell to governments and their schools and universities in partnership with a plethora of social organizations. The intense nationalism accompanying World War I, World War II, and the Cold War furthered a general consensus among social scientists that local communities and allegiances would gradually give way to national societies.

Throughout much of the twentieth century, the territorial trap relegated border research to subfields like political anthropology, political geography, or regional politics, economics, and sociology. Nevertheless, beginning slowly in the 1960s and into the 1970s with decolonization, gaining momentum through the 1980s with the waning of Soviet power, and booming in the 1990s with the emergence of new states following the collapse of communism, the topic of borders gained prominence in academic research as scholars questioned the overt emphasis on "security" rather than "opportunity." Border research gained even greater relevance through the first decades of the twenty-first century because of increased economic integration, migration flows, global environmental awareness, and socioeconomic inequality, among other issues. Today, mounting concern about the disparities between haves and have-nots draws many to question the morality of linking opportunity for human development and overall well-being to the largely arbitrary lines that define state sovereignty.

The contradictory and controversial nature of contemporary borders has provided fertile ground for research, including the establishment of an array of research centers and a growing community of border scholars. The proliferation of contested borders and border-related issues has been accompanied by, somewhat ironically or perhaps even as a catalyst for, a growing sense that borders are or should diminish in importance. Greater mobility, the development of regional and supranational institutions, and global economic integration may well engender a re-territorializing of national identity, sovereignty, and economies. It is fair to say that the notion of an end of geography—connoting a world of flows erasing the world of places and ultimately giving rise to a homogenized, borderless landscape—has been largely discounted. In its place, the sense that the border is everywhere has gained broad currency. Technology enables instantaneous connections around the globe, while individuals can be subject to unprecedented monitoring and surveillance. In the future, the scale of community and the capacity to exercise power may not require fixed links to geography.

But far from sounding the death knell for borders as the default organizing unit for global political space, many academics and analysts project a scaling up to larger continental or supranational geopolitical structures that remain demarcated by borders. Others foresee a scaling down as states fragment into regional territorial structures with overlapping and less rigidly defined local identities and concepts of sovereignty. Both alternatives entail new functions for territoriality and borders rather than their abolition. The effects of ethnic or national populism and globalization are so wide-ranging that a singular effect on human spatiality is unlikely. Indeed, history reveals alternating patterns of border upheaval and stability, permeability and rigidity, as well as cosmopolitanism and isolationism.

The collapse of the bipolar international system along with its figurative and literal border landmarks—the Iron Curtain and the Berlin Wall—resulted in renewed border instability across the former communist realm. The Soviet Union disintegrated into fifteen new states, and Yugoslavia and Czechoslovakia likewise broke apart into new sovereign states during the 1990s. The repercussions are still unfolding through the early twenty-first century. Ethnic Albanians in Kosovo proclaimed their nation-state status in 2008, while Russia invaded Georgia in support of independence for South Ossetians and Abkhazians that same year. Armenia and Azerbaijan engaged in bloody clashes over the disputed Nagorno-Karabakh region in 2020. Russia would go on to annex Crimea in 2014 and support Russian-speaking secessionist movements in the Donbas region as precursors to its full-scale invasion of Ukraine in 2022.

Interestingly, countries as diverse as China and Spain have been reluctant to recognize new claims to sovereignty for fear of fueling movements demanding national self-determination among their own dissatisfied minority groups—Tibetan and Uyghur, Basque and Catalan respectively. The United Kingdom's withdrawal from the European Union—the so-called Brexit—raises numerous border questions ranging from Scottish independence to the evolving nexus between the Republic of Ireland and Northern Ireland. The fracturing of empires, cobbled-together federated states, and regional blocks, however, covers just some of the recent border-related headlines. Ironically, ongoing efforts among various Muslim movements to establish some type of caliphate and Turkey's role as a buffer territory between the European Union and the Middle East, institutionalized through the 2015 EU-Turkey Joint Action Plan, highlight actions favoring different forms of consolidation and cooperation.

There is no single trend toward either integration or fragmentation. Some parts of the world are moving toward greater connectivity and integration, while other regions move in different directions. Borders remain significant within the flows of globalization, because borders provide the "trans"—that is to cross, breach, or span—in processes of transnationalism, transmigration, and transgression. Though potentially changing through novel connections generated by various types of cross-border flows and communities, territoriality remains ubiquitous and unlikely to fade any time soon.

We are, nevertheless, living through a transition of human spatiality, territoriality, and borders. Contradictory border processes may trigger broad cultural anxieties as societies and governments adapt to evolving and complex spatial realities. Beyond the effectiveness of borders in striking a balance between security and opportunity, the ethics of borders and bordering have been foremost topics in border scholarship since the first edition of this book was published over a decade ago. One faction of scholars and activists calls for "open borders"—nation-states would remain, but the function of borders would evolve to allow the free movement of people, goods, and information, and the overall sharing of opportunity. Another faction calls for "no borders"—the nation-state system would essentially end as the world adopted common political, socioeconomic, and cultural frameworks on a global scale. Some formulate this evolution in terms of neoliberalism, while for others it is rooted in socialism, and for still others it would portend anarchy.

External state borders are central to a variety of issues, including tourism, labor rights, environmental conservation, and crime, but

2. Protestors in London call for the European Union to open its borders.

new socioeconomic and political linkages are also producing
unique forms of bordering and alternative spatial realities across
substate scales and jurisdictions. Voting districts, census tracts,
municipal boundaries, and any number of other administrative
divisions of space, along with unofficial boundaries of
socioeconomic and cultural differences, constitute tangible
landscapes of status, authority, and power. These administrative
hierarchies and their varied spatialities play important roles in
shaping and contextualizing individual and group identity. Our
relative position within these hierarchies—this side of the tracks
or that, this neighborhood or that, this city or that, this country or
that—shapes our sense of self, belonging, and overall life
opportunities. The aspirations, prospects, and perceptions of
people are conditioned, in large part, by the varied degrees of
power vested within those territorial units. A variety of scholars
and activists regard this "birthright lottery" as establishing a
global apartheid in which borders protect the privilege of some
and restrict others to places of diminished resources and
opportunity.

Ultimately, those calling for open borders or no borders face several substantive challenges, not only from the inertia of the nation-state system in combination with capitalism, but also on a popular front. Many regard borders as preserving their hard-won and cherished communities of culture, traditions, and standards of living. Moreover, while most people would feel relatively comfortable revising a census tract or local park boundary, the lines dividing the colorful collage of countries on world maps convey an air of sanctity. Different perceptions of the significance and permanence of geographic boundaries are not accidental. International borders have been purposely constructed and represented to appear as though they derive from some higher logic. They are, however, no more natural or logical than obviously contrived school zones or electoral districts.

Whether based on seemingly objective criteria, such as rivers or lines of latitude, or appearing convoluted and artificial, all borders reflect the capriciousness of human beliefs, assumptions, biases, and agendas. This does not mean that borders lack power and significance. Territorial disputes and competing border practices are often so intractable because individuals or groups are supremely confident in the justness of their respective positions. As such, every geographic boundary is a symbolic representation and practical embodiment of human territoriality and by extension identity. This has been true since the first group of humans referred to some piece of land as "ours" as opposed to "theirs."

Today, many governments struggle to reconcile the management of cross-border economic exchange, competitive international security environments, surges of distraught migrants, and dynamic environmental issues with their inclination to prioritize the economic and security demands of their constituencies. While most religions share the foundational belief with secular cosmopolitanism that morality should extend beyond the confines of any specific identity and territory, it nevertheless seems every proposal to "soften" borders is met with equal efforts to "harden" borders and protect "what's ours." Borders are a technology of social control that, in some cases and for some people, is benign or beneficial offering sanctuary and security; while in other cases and for different people, such control is oppressive or violent, constituting a prison to those within or a fortress against those outside. Thus, rather than holding an intrinsic positive or negative value, a border's utility and morality, like most technologies, depends on how it is used.

That said, any strengthening of border control must contend with efforts to circumvent those restrictions. Every desire to block is met with an impulse to breach. Neither holds an absolute moral high ground. The free flow of cyberspace might facilitate novel human connections by reducing the friction of distance, but it may simultaneously provide a venue for confrontation, discrimination, and hatred. The extent to which humans have the right to draw and enforce borders to regulate flows into and out of a territory is inherently linked to possibilities for defining normative responsibility for that territory. In this transformative period of shifting spatiality and territoriality, the balance between those different ethical concerns rightly constitutes a driving force moving border research forward.

A very short introduction to borders

Borders are integral components of human activity and organization. As such, we are compelled to deepen our understanding of their varied potential to serve as areas of opportunity and insecurity, zones of contact and conflict, sites of cooperation and competition, places of ambivalent identities and aggressive assertions of difference. These dichotomies may change over time and across space, but more interestingly, they often coexist. We must come to terms with how borders structure our lives, while simultaneously lessening their power to perpetuate resentment and indifference. We must find a way to harness their ability to catalyze belonging and solidarity while diminishing their propensity for exclusion and the creation of "others." We must navigate the increasingly mobile, malleable, and multi-scalar nature of territoriality and borders in a variety of social settings. We must tackle ethical questions: For whom are borders constructed? By whom? And to what ends?

Borders require further study both from the top down and from the bottom up, from the state scale and from the local scale, as they are among the most ubiquitous geographic realities in our lives. Ultimately, the lived experience of borders reminds us that recognizing their opacity, permeability, and rigidity is as important as their transparency, porousness, and mobility. How do borders evolve in function, utility, and morality? How do sites of cooperation become sites of contestation, and vice versa? Like other Very Short Introductions, this book cannot provide a comprehensive treatment of such a diverse and complex topic as borders. It is for that reason that borders are and will remain such important factors shaping our understanding of the world and our place in it.

Chapter 2
Borders and territory
in the ancient world

Modern political maps confer a sense of permanence for the world's states, and by extension their territories and borders. Indeed, a major objective of the modern social sciences has been to create historical lineages bestowing prehistorical origins to contemporary political entities, such as China, France, and Iran, and their sovereign territories. Yet the modern political map and its underlying territorial assumptions are relatively recent developments. This is not to suggest that premodern societies and polities lacked notions of territoriality or borders; rather, they tended to conceptualize and organize space in ways that differ from contemporary expectations and norms.

Territoriality among hunter-gatherers

The earliest humans lived in small bands of nomadic foragers—known as hunter-gatherers—that moved about based on changing seasons and environmental conditions that affected the availability of food and other resources. Given their nomadic way of life, one might assume those early peoples lacked notions of borders, territoriality, and land ownership. Yet research on contemporary hunter-gatherer groups strongly suggests that their prehistoric counterparts practiced rather elaborate territorial strategies. Instead of wandering aimlessly, hunter-gatherer groups normally maneuvered across relatively stable local or regional foraging ranges, maybe better described as networks of foraging sites. These networks were shaped by extended kinship or alliance relationships, religious beliefs, and ecological conditions. Yet notions of ownership and norms regarding access to territory and resources varied considerably.

Some groups invested considerable energy in claiming and maintaining exclusive access to specific hunting and foraging areas. The Tsimshian, Coast Salish, and many other groups in North America's Pacific Northwest developed intricate social systems governing ownership of and access to resource sites. Punishment for trespassing could be severe, but property-owning lineages were expected to share their bounty through ritual celebrations. The Veddah peoples of Sri Lanka also marked and rigorously safeguarded specific territories. Each band was expected to be self-sufficient within its own hunting area, and trespassing was strictly prohibited in most circumstances. In areas lacking an obvious natural landmark, the Veddahs would carve symbols into tree trunks to mark borders. Although these and other hunter-gatherer groups may have shared a notion of exclusive territoriality, the concept of ownership resided with individual households in some groups, while broader familial lineages owned land in common in others.

Other hunter-gatherer cultures adopted alternative territorial strategies for land and resource management that emphasized social cohesion and reciprocity. The !Kung bands of southern Africa had rather vague borders for their foraging ranges, often coinciding with natural landmarks, and made little effort to deny access to other bands. Resources were not considered to be owned until gathered and should be shared equally once collected. Bands had considerable freedom to access resources in neighboring ranges, especially with those bands sharing kinship ties. Such visits were frequent and generally welcome, so long as the visiting group first sought permission and shared some portion of the collected resources with the host band. Similar hospitality was expected in return. Many Aboriginal groups in Australia followed a comparable system where fluid membership and kinship ties also encouraged territorial and resource sharing between neighboring bands. Each band may have had a specific foraging network, but these ranges were not considered exclusive and often overlapped with neighboring groups. In other cases, foraging ranges were separated by buffer areas, which were used infrequently and apparently not claimed by any group.

Territorial practices among hunter-gatherer groups exhibited a great deal of variability, flexibility, and adaptability as bands merged, divided, or sought out new lands in response to population changes, resource availability, or personal disagreements. Although it is tempting to think of ancient hunter-gatherer groups as nonterritorial since clearly defined borders and territories are difficult to discern, these early groups most likely featured complex spatial strategies for regulating membership, territory, and resources that intermingled notions of exclusivity and reciprocity shaped by animistic religious practices, kinship ties, resource abundance, and environmental conditions.

State formation in antiquity

Starting around the twelfth millennium BCE, hunter-gatherer bands gradually adopted sedentary lifestyles in farming communities. This so-called Neolithic Revolution was triggered by the domestication of plants and animals, which first occurred in an area of productive farmland roughly stretching across modern-day Iraq, Turkey, Syria, and Lebanon known as the Fertile Crescent. Settled farming societies arose independently in parts of Africa, Eurasia, and the Americas by the third millennium BCE. In addition to permanent settlements, this shift from hunting-gathering to farming had far-reaching consequences for human history, including the development of writing, irrigation, architecture, government, and socioeconomic specialization. The shift also facilitated the emergence of new socioeconomic and political organizations of territory. Modest villages and later city-states ruled by local chiefs and kings gradually developed, especially along fertile river valleys.

The Sumerians were the earliest known group to establish a city-state culture. By around 4000 BCE, the Sumerians began to coalesce into a few dozen substantial city-states scattered along the Tigris-Euphrates river basin. A city-state is a sovereign polity encompassing a relatively small area consisting of an urban core and surrounding farmland. In the case of Sumer, each city-state was independent and ruled by a king, who represented the city's patron deity. These city-states relied on the agricultural surpluses produced in surrounding fields to sustain the urban population. If a king wished to expand his city-state and thereby increase his power and wealth, he generally needed to acquire more farmland, which normally entailed expanding irrigation canals to form new fields or annexing established fields from neighboring city-states. Plundering other city-states was also a common occurrence.

Given that context, it is not surprising that Sumerian city-states seemed in near constant conflict with their neighbors. Some kings were quite successful and succeeded in consolidating numerous city-states into kingdoms. Yet the earliest conquests in Sumer seemed more likely to result in plunder and tributary payments from the defeated city-state rather than outright territorial annexation, and even these tributary arrangements appeared fleeting. Evidence for this comes from *stelae*—or commemorative stone slabs, plural of *stele*—which ancient rulers commissioned to celebrate their accomplishments. Stelae were commonly used to commemorate important events like a military victory, building dedication, or a ruler's accession to the throne, but some were also used to establish territorial claims and boundaries. Some marked internal administrative boundaries, while others represented external borders between neighboring city-states.

In ancient Egypt, for instance, the pharaoh Akhenaten marked the limits of his new capital city at Amarna with a series of stelae. In addition to indicating the city's limits, the stelae were also inscribed with religious dedications, which distinguished the sacred precinct of the capital from the rest of the kingdom. Other pharaohs used stelae to mark the outer limits of their realm, such as Senusret III who erected stelae to designate Egypt's southern border with Nubia. Yet the stelae were not intended to mark a clearly defined border line. Instead, they were located inside frontier fortresses and made general claims over the surrounding territories rather than marking a sharp dividing line between sovereign territories. A single border line in the modern sense did not exist. Rather, the inscriptions on these stelae and nearby statues suggest the fortresses served more as "checkpoints" where government officials could regulate the movement of people and goods along the Nile, a vital travel corridor. Stelae served similar functions for the Olmec and Mayan civilizations in Central America. These stelae have provided invaluable evidence for understanding these ancient cultures and their notions of territory and borders.

The so-called Stele of the Vultures and other ancient inscriptions record the earliest known border conflict, specifically a dispute over farmland between the prominent Sumerian city-states of Lagash and Umma during the twenty-fifth century BCE. The top Sumerian deity Enlil had originally set the border between Lagash and Umma, but the two cities disputed the actual location. Acting as some sort of arbitrator or even divine intermediary, Mesalim, the king of the city-state Kish, later surveyed the border and erected a stele inscribed with his decision. Despite this, subsequent rulers in Lagash and Umma quarreled incessantly over the fields and water rights in the area. Eventually, the ruler of Umma removed the stele and seized all the land. Several years later, Eannatum, ruler of Lagash, struck back and defeated Umma. Eannatum restored the original boundary stele and dug an irrigation canal to mark the border. He also erected a second stele outlining a harsh peace treaty imposed upon Umma. The treaty specified which fields belonged to Lagash and Umma, obliged Umma to maintain a buffer zone on its side of the border, and required payments of tribute to Lagash. Finally, Eannatum sought to affirm the border's sacred sanction by building shrines to Enlil and other deities. These stelae and shrines marked only a portion of the border, but they contained inscriptions describing its entire course.

The city-state would remain a prominent form of territorial and political organization for centuries, but broader trends were already emerging that enabled the formation of larger and more centralized structures. As agricultural yields improved, city-states were able to grow demographically, specialize socioeconomically, and support expanded government bureaucracies and standing militaries. These changes made it more feasible for ambitious rulers to conquer and annex neighboring city-states, rather than simply plundering and demanding payments of tribute.

Again, the ancient Fertile Crescent provides the earliest known example in Sargon the Great, ruler of Kish. Although ruling dynasties in antiquity were continually being established and overthrown, Sargon managed to create and sustain an expansive and relatively centralized system of territorial rule. Following a series of military victories, Sargon established the Akkadian Empire, which encompassed most of the formerly independent city-states of Mesopotamia and adjacent regions during much of the twenty-fourth and twenty-third centuries BCE. It is generally recognized as the world's first empire.

An empire is a sovereign political entity, usually governed by a hereditary monarch, and spanning several different regions and peoples that had not traditionally been under a single ruler. Empires require more complex, multilayered, and extensive administrative structures to govern effectively because of their greater size and diversity compared with traditional city-states. This often entailed the delegation of authority to lower-level governors and administrators who acted on behalf of the central authority. In the case of the Akkadian Empire, Sargon and his successors appointed governors, or *énsi*, to administer the areas beyond the capital city district. These provincial governors were usually selected from the Akkadian elite to ensure loyalty to the central imperial authority.

Despite their apparent sophistication, the Akkadian Empire and its successors in antiquity proved fragile. Internally, the elites of conquered city-states commonly rebelled to reclaim their time-honored and cherished independence, as was the case for Sumerian cities, which repeatedly rebelled against their Akkadian masters. Additionally, the king could face jealous rivals within the capital district or even from within the royal family. Externally, the threat of invasion may have spurred neighboring city-states toward greater cooperation and even centralization of political power to defend against Akkadian or Sumerian attacks. This may have happened among the Elamites in what is now southwestern Iran. Prosperity within empires likely made them more tempting targets for neighboring city-states and empires, as well as nomadic groups. An invasion by the Gutians, nomads from the northern parts of modern-day Iran, eventually brought an end to the Akkadian Empire.

This summary highlights the three main modes of political-territorial organization that dominated much of history: nomadic pastoralists (i.e., Gutians), city-states (i.e., Lagash), and empires (i.e., Akkadian). A fourth possibility was some type of federation encompassing various tribes, city-states, and empires banding together to face a common opponent, although these types of arrangements were generally fleeting. Over the following centuries, successive empires were established, expanded, and collapsed, including the Assyrian, Persian, and Roman Empires. City-state cultures thrived among the Greeks, Philistines, and Phoenicians. Nomadic and seminomadic groups, like the Amorites,

Aramaeans, and Hyksos, continually plundered and occasionally conquered these city-states and empires. It is important to avoid interpreting this as a predetermined progression, or "scaling up," from simple marauding bands to modest city-states to larger complex empires. Rather, the ancient world's political scene fluctuated back and forth between these varied political frameworks in response to changing environmental conditions, agricultural productivity, trade routes, military technologies, and the relative strength of neighboring groups. Despite changing contexts, each of these political frameworks employed some sort of spatial strategy or system to manage territory, people, and resources.

Nomadic pastoralists

Ancient nomadic pastoralists may resemble prehistoric hunter-gatherers, but that impression is superficial and misleading. Nomadic pastoralism developed as a system of livestock-rearing in areas with limited farming potential, such as semi-arid regions of the Central Asian steppe or Africa's Sahel. Although little is known about the territorial organization among these groups in antiquity, such as the Scythians or Huns, much can be inferred from groups like the Kazakhs or Mongols, who retained nomadic pastoral lifestyles into recent times. Although they generally did not conceive of individual land ownership or that land had any intrinsic value, their reliance on animal husbandry shaped their political and territorial organization. Among these pastoral groups, territorial access was framed around the movement of livestock between seasonal grazing ranges, which were generally conceived as belonging to specific lineages or kinship groups.

The Eurasian steppes had been home to various nomadic pastoralists for centuries. These rather decentralized groups occasionally came together to launch plundering raids or repel outsiders encroaching upon commonly held grazing areas. These pastoralists posed a recurring threat to sedentary cultures but, conversely, they also constituted important trading partners and at times powerful military allies. Occasionally, powerful chiefs were able to assemble confederacies of pastoralists strong enough to conquer their nomadic and sedentary neighbors. The empire established by Mongol leader Chinggis Khan (Genghis Khan) was the largest of these and eventually covered much of the Eurasian landmass by the late thirteenth century CE. Chinggis Khan and his successors developed a more centralized administrative structure to govern this immense territory. While traditional land-use practices remained under the supervision of local tribal leaders, the khans appointed governors to oversee areas beyond the Mongol homeland. Instead of well-defined external borders, the Mongol Empire linked these far-flung administrative centers through an extensive network of roads, relay stations, and outposts known as the *Yam*, which served as arteries for information, commerce, migration, and military deployments stretching across Eurasia.

Although lacking rigid border lines, nomadic pastoralists possessed sophisticated notions of territoriality and boundaries. Walled or fenced burial sites exist in several pastoral cultures. These groups, however, tended to apply more fluid notions of territoriality at larger scales. For most pastoralists across Eurasia, the idea of borders related more to ecological zones or simply distance. Archeological evidence of tool and other material exchanges, such as furs, hides, crafts, and fabrics, suggests that the Bronze and Iron Ages saw a great deal of cultural mixing between forager and pastoralist cultures, especially at ecological transition zones. Long before the "Silk Road" spanned the plains of Central Asia, a "fur route" linked the steppe pastoralists to distant lands and peoples of the north. Although firm borders and direct claims of ownership appear less prominent among pastoralists, distinct markers of territorial occupancy and/or authority are found in burial sites, funerary monuments, and *balbals*—sometimes known as deer stones—scattered across Eurasia. The geopolitical and social role that these monuments played continues to be hotly debated, though. The extent to which political organization among these groups required any territorial demarcation remains unclear. Pastoral economies have proven viable both within imperial state structures and without any political hierarchy at all. From East Africa to Central Asia, one finds pastoralists spurning central political authorities throughout history. The claim that pastoralists lacked a sense of territoriality is spurious, but it is fair to say that their notions of territoriality were far more fluid than those of sedentary groups. It was, in many ways, the fluidity of pastoral territoriality that catalyzed specific architectural and sociopolitical adaptations among expanding sedentary populations.

City-states

City-states were modest in area and population, typically ranging from 250 to 2,500 square miles (650–6,500 km²) and encompassing 5,000 to 25,000 inhabitants. In rare cases, they grew far larger, such as ancient Athens or pre-Columbian Tenochtitlan. Scholars have generally regarded city-states as either dead ends or transitional forms in the political evolution from simple chiefdoms to complex empires and eventually modern states. These interpretations reflect the uncritical projection of modern-day territorial assumptions and practices back through history and ignore the fact that city-states flourished as the dominant political organization in various parts of the world over long periods of time. In addition to their cultural and technological achievements, the Sumerian city-states managed to persist for nearly two millennia. Even when incorporated into the Akkadian Empire, Sumerian cities revolted often and eventually regained their independence. This is testament to the city-state's enduring appeal and viability in the region. City-states also proved long-lasting in other parts of the world, as demonstrated by those of the Nahuatl in the Valley of Mexico, the Phoenicians throughout the Mediterranean, and the Yoruba in West Africa.

Given their small size, these city-states may appear rather unsophisticated in terms of territorial organization and governance, but that was not the case. Rather than isolated, self-sufficient entities, ancient city-states tended to exist in spatial clusters engaged in intensive exchange and interaction. These city-state clusters functioned as hierarchical yet flexible networks of political, socioeconomic, and technological cooperation and competition. These networks provided a major incentive for city-states to adopt modes of territorial organization that maximized opportunities while minimizing threats. In practice, this often meant facilitating mutually beneficial economic, technological, and cultural exchanges with neighboring city-states, while maintaining a vigorous defense against external threats. Yet there were significant variations in the time, resources, and energy invested in border creation and maintenance, as well as which types of borders assumed priority.

The city-states—or *poleis*, plural of *polis*—of ancient Greece are among the best studied of the early city-state cultures. Much is known about the culture, politics, and economies of Greek city-states, but less effort has been made to understand how ancient Greeks organized territory. In fact, scholarly attention is largely confined to within the city's fortifications. This perspective is illustrated by the tendency to represent Greek city-states as mere dots without mapping out their territories. Yet there are numerous instances of disagreements over territory leading to conflict between city-states. It appears likely that most city-states, especially larger ones, placed stone markers—known as *horoi*—at locations where travel routes crossed their outermost territories. Religious sanctuaries were often located nearby as a way of providing divine justification for the city-state's territory. Travelers often marked border crossings by offering a sacrifice to the gods. Neutral zones commonly separated the borders of neighboring city-states. By mutual agreement, this unclaimed territory was open for livestock grazing, but the establishment of fields or settlements was prohibited. The "border" wars among Greek city-states appear just as likely to center on disputes over the usage of this neutral territory rather than the actual location of the border. This neutral zone and the border's religious connotations help explain why relatively little territory was transferred between poleis as a result of warfare. Defeated city-states were more likely to be forced to ally with the victor or pay tribute than lose territory.

Unlike the Greek polis, the classical Mayan city-state—or *ahawlel*—generally lacked exterior walls, but considerable effort went into marking the borders between city-state territories. The process of border creation involved elaborate ceremonies after which participants would walk the boundary, marking it as they went. The process of dividing farmland within each city-state's territory was equally important. Each village had its own farming territories, which were organized into parcels. The Maya did not buy or sell land, so parcels were passed down through familial lineages, which maintained the borders of their farmlands through rituals that combined walking the border with ancestor worship.

Clearly some city-state cultures devoted significant time, labor, and resources to marking their territories, but this was not universal. The Malay city-states—or *negeri*—of the fifteenth and sixteenth centuries seem to have done little to mark or maintain borders. These Malay states were structured around river systems with the main urban center located at the river's mouth. The city's hinterland consisted of loosely controlled villages scattered upstream along the river and its tributaries. Because the territory away from the rivers was generally mountainous jungle and therefore poor for farming, it was sparsely populated and difficult to traverse. As a result, the main city at the river's mouth could regulate the flow of resources, goods, and people into and out of the river basin without having to maintain extensive inland borders. It appears that the coastal centers did not even bother to implement direct rule over their hinterlands, preferring instead to leave local chiefs in charge. Instead of the compact and contiguous territories seen on modern maps, the territories of the Malay city-states are better imagined as networks of coastal and riverside communities resembling a tree. The core coastal city-state was located at the base of the trunk while inland villages were scattered along the upper branches. Any trade between river basins would almost always mean passing through their respective city-state cores.

Several factors explain these varied border strategies, including differences in environmental conditions, settlement and population densities, and agricultural and economic practices. Those areas beyond Greek city-states were often productive grazing lands, but lands away from rivers were of little economic use for Malay city-states. There was less incentive to mark off land deemed unproductive. Greek colonies in North Africa placed markers along the coast to separate neighboring territories but, seeing little economic value in the desert interior, made little effort to extend borders inland from the coastal markers. A similar pattern manifested in the seventeenth and eighteenth centuries, as Europeans marked the coasts of Southeast Asia but largely left the interior jungles without demarcation. The Greeks were also content to leave neutral zones between territories, but the Mayans appear to have assigned all the territory to one city-state or another. This might be explained by the Greeks' need for grazing lands, which could be used in common with neighboring city-states, while the Maya practiced intensive agriculture, which required clearer partitions of land. In contrast to both, the Malay city-states relied more on the sea and rivers for their food, as well as for trade. These different environmental, agricultural, and economic contexts fostered the development of different territorial strategies. Despite the apparent differences in their attitudes toward borders, the Greek *polis*, the Mayan *ahawlel*, and the Malay *negeri* all employed territorial strategies that harnessed the labor, resources, and production in their hinterlands for the benefit of the urban core.

Empires

The empire was the third political framework common in antiquity. Empires usually formed when one city-state, or less commonly a nomadic group, succeeded in establishing political and economic dominance over an expansive, multiethnic territory beyond its traditional homeland. Most empires were governed by hereditary monarchs whose legitimacy was based on notions of religious or universal kingship; that is, the monarch claimed a divine mandate to govern. As a result, empires tended to pursue expansionist policies aimed at subjugating neighbors, whether they were nomadic groups, city-states, or other empires. It is easy to think of empires as simply successful city-states that grew very large, similar to the transformation of the Babylonian or Roman city-states into their respective empires. Yet empires required more expansive and hierarchical territorial structures to govern their varied peoples and regions effectively.

In general, early empires can be grouped into two main categories based on whether territorial control was exercised through direct or indirect means. Some empires developed relatively centralized systems that concentrated authority in the imperial capital. The territory beyond the imperial core was divided into provinces headed by regional governors. These governors, normally drawn from elite families based in the imperial core, were responsible for maintaining law and order, implementing policies from the central government, and perhaps most importantly, overseeing the collection of taxes and resources to be shipped to the imperial capital. The Persian Empire was divided into approximately two dozen provinces each headed by a governor—or *satrap*—who ruled on behalf of the emperor. Similar arrangements have been documented in the Egyptian, Inca Empires, and Roman.

Other empires relied on less centralized means of control. The Aztec Empire was essentially a city-state, Tenochtitlan, that succeeded in dominating the other city-states in central Mexico. Aztec rulers did not directly annex those city-state territories but rather generally allowed local kings to remain in place so long as they provided military service and sufficient tribute to the central government. Local kings retained considerable authority over local matters and their positions were hereditary, unlike the governorships in more centralized empires. Although they may appear quite different, these forms of rule shared a great deal. They were all intended to channel tribute, labor, and resources to support and benefit the imperial core. The possibility that local kings or governors might try to break free was also a constant problem for both and often a more serious threat to the empire's survival than hostile neighbors.

Regardless of their internal administration, empires sought some way to structure their territory. The Roman Empire appears to embody the ideal of a highly centralized empire with well-defined borders. The surviving portions of Rome's border defenses reinforce this perception for modern viewers. Remnants like Hadrian's Wall—an extensive network of forts, towers, and walls in northern England—suggest clear and definite limits of Roman territorial control. Yet the impression of Roman soldiers standing guard along the wall against barbarians on the other side is misleading. Rather than marking the limit of Roman control, the Romans used Hadrian's Wall to project their authority well to the north. In this sense, the Roman use of Hadrian's Wall was similar to how imperial dynasties used the Great Wall of China. Both frontier fortifications suggest clearly defined limits to imperial authority but rarely marked an exact border between their respective empires and their northern neighbors with whom they both traded and clashed. Archeological evidence of settlement patterns suggests that pastoralists and sedentary farmers were common on both sides of China's Great Wall, much like material remnants of Celtic peoples and Roman settlers span Hadrian's Wall. In practice, these walls provided imperial troops a platform for controlling territory and regulating the movement of people and goods on both sides. The

3. Ancient fortifications, such as the Great Wall of China, rarely marked a rigid political border.

ability of Chinese and Roman forces to actually achieve these goals varied over time. The fact that successive Chinese dynasties were compelled to construct new walls over so many centuries and that Rome eventually abandoned Hadrian's Wall reflects the challenge of maintaining stable frontiers.

Although impressive structures like Hadrian's Wall or the Great Wall suggest clear linear borders, it is more helpful to think of early empires as relatively fluid, indeterminate territories. Research on the Assyrian Empire highlights the degree to which premodern states pursued different territorial strategies for different sections of their frontiers. Assyrian rulers appeared to adapt frontier strategies based on an area's relative economic or strategic importance. In areas deemed important, the Assyrians took direct control by establishing new administrative centers. The forced importation of agricultural colonists to newly conquered regions increased the area's productivity and ensured a more loyal populace. In areas deemed less important, the Assyrians seemed content to maintain neutral zones occupied by village-level chiefdoms, which posed no security threat. It also appears that Assyrian control in its outer provinces was patchy, with imperial power firmly established in some areas but apparently absent in others. Instead of a contiguous territory, the Assyrian Empire is better described as a network of dispersed pockets of imperial control interspersed with areas lacking significant political organization. Similar variation has been found in the frontier policies of the Aztec and Sassanid Empires. Depending on differences in the natural terrain, strength of opposing forces, and economic importance, some frontiers were administered indirectly through local rulers who pledged loyalty and tribute to the empire, other provinces were placed under direct imperial control, and still other stretches of frontier were basically left open.

Ancient polities as flexible territorial structures

Ancient nomadic pastoral bands, city-states, and empires employed a range of beliefs and practices for organizing space and bordering territory. Given differing political, socioeconomic, environmental, and technological contexts, it should not be surprising that groups in different parts of the world and in different times developed varied approaches to territory, borders, and governance. These approaches were more nuanced and flexible than commonly assumed. Several scholars have argued that these ancient governments were more focused on controlling the movement of people than controlling actual territory since labor supply was the key limiting factor in providing sufficient food, resources, soldiers, bureaucrats, and servants to support the ruling elite. If correct, this helps explain why many ancient governments were generally satisfied with relatively vague borders.

However, modern maps tend to depict these early pastoral groups as nonterritorial, city-states as mere dots, and empires as clearly bounded political territories. These cartographic conventions have less to do with historical evidence than with the projection of contemporary assumptions of the modern world back on to previous centuries. Such a simplistic perspective leads to two general misconceptions regarding state formation, territoriality, and borders in antiquity. The first interprets history as an inexorable progression from hunter-gatherers to farming villages or pastoralists to city-states to eventually empires, or from small and simple to large and complex. Yet there are numerous cases of empires fracturing into smaller-scale polities or pastoralists building empires without passing through the city-state "stage." These and other scenarios refute notions of linear political evolution or social progression. A second misconception presumes that ancient polities possessed uniform control over territories marked by clear borders. Although early nomadic pastoralists, city-states, and empires may have claimed absolute authority over a given area, a variety of strategies for political-territorial control were used over the centuries that provided governments varying degrees of authority and integration within their nominal territories.

Internally, these ancient governments may have functioned more like flexible, patchwork networks of territorial control interlinked through strategic transportation corridors. Externally, territorial borders were more like transitional frontier zones containing a fluid mixture of imperial strongholds, client states, opposing forces, and areas of indeterminacy, instead of sharp lines separating clear spaces of sovereignty. Areas with limited accessibility, either through sheer distance from the central government or natural obstacles like mountains, deserts, or jungles, often straddled the distinction between internal and external and provided refuge for various "marginal" groups to operate autonomously, despite occupying territory nominally claimed by some higher sovereign ruler. The desire to eliminate these territorial gaps and inconsistencies in the exercise of state power provided a powerful impetus for new territorial and bordering strategies that ultimately contributed to the formation of the modern state system.

Chapter 3
The modern state system

In the ancient world, territorial control was often exercised without clearly defined borders. Yet the varied and flexible approaches found throughout history gradually gave way to more rigid notions of borders, territory, and sovereignty within what is commonly known as the modern state system. Largely taken for granted today, the modern state system refers to political entities—commonly called countries—that recognize each other as possessing absolute sovereignty over a bordered territory, and the people, resources, things, and generally everything else within that territory. The development of this state system was a gradual and, in many ways, incomplete process. The modern state system first emerged in Europe and spread to the rest of the world through colonialism, intimidation, and imitation. Understanding the emergence of the state system provides a foundation for understanding contemporary debates about the role of borders and territory in our increasingly globalized world.

The origins of modern states

The modern state system traces its origins to complex socioeconomic, political, and technological changes that began coalescing in Western Europe as early as the eleventh century. Some familiar names could be found on Europe's map at that time, like France and England, but those entities were very different than their modern successors. Instead of distinct sovereign states, medieval Europe was organized around what was later called the feudal system. Feudalism was a form of political organization encompassing a complex framework of privileges and obligations between lords and vassals. As the Carolingian Empire declined and fragmented during the ninth century, the Frankish kings struggled to pay their military commanders. Instead of monetary payment, the kings appointed those commanders as vassals with the right to govern a portion of the king's land—commonly called a fief—and control its economy as a form of payment. In return, the vassals pledged loyalty and military service to the king. As a result, feudalism was based more on personal oaths and responsibilities than on rigid territorial organization. Kings and vassals were bound by their personal commitments regardless of their territorial location.

Feudalism seemed to offer a clear hierarchy, with kings and leading nobles at the top, knights and lesser nobles in the middle, and peasants, serfs, and slaves at the bottom. Initially, the king retained ultimate ownership of the territory and could reassign it if the vassal was disloyal, died, or simply fell out of favor. However, as central authority weakened even further, vassals gradually gained hereditary title to the lands and became largely independent, and occasionally competitors or even enemies of their nominal lord. Marriages and land transactions between noble families, elaborate inheritance customs, and military conquests further complicated the situation, resulting in a system of decentralized political authority, discontiguous territories, and overlapping jurisdictions. The dukes of Burgundy possessed large land holdings within the kingdom of France, but they also held considerable territories in the Holy Roman Empire. On top of this, the Roman Catholic Church claimed a degree of universal authority. Therefore, the dukes owed some form of allegiance to the French kings, German emperors, and Catholic popes simultaneously. Given this confused structure, precise territorial borders were not necessarily needed or helpful, so long as taxes were collected, services rendered, and oaths fulfilled, especially in sparsely populated regions.

Ironically, a political structure based on state sovereignty with clearly marked territories originated from a feudal system characterized by patchwork jurisdictions, overlapping allegiances, and vague borders. Scholars debate the exact causes and timeline, but there is general agreement that the modern state system began to form by the late Middle Ages as centralized governments exercised increasing political and economic control over defined territories. The reasons for this shift are complex but generally involve a series of economic changes beginning around 1000 CE, including improved agricultural yields, increased socioeconomic specialization, monetary exchange, and long-distance trade. These changes fueled the growth of cities and increasingly powerful new classes of skilled workers, merchants, traders, and financiers.

The growing power of cities disrupted established feudal relationships and provided opportunities for new political arrangements. Cities generally shared the same objectives of political security and stability that would foster economic growth and trade. Kings and feudal nobles hoped to harness the growing wealth of these cities in their struggles with each other. Yet differences in the relative strength of cities, nobles, and monarchs across Europe led to different political outcomes, namely the emergence of city-leagues, city-states, and territorial states. This redistribution of power was accompanied by new approaches to territorial organization that gradually replaced the decentralized logic of feudalism and the universalistic claims of popes and emperors.

In Germany, cities were not strong enough to act individually, and the German kings were weakened by their failed attempts to incorporate Italy into the Holy Roman Empire. As a result, cities banded together into city-leagues to defend their interests against the nobles. These leagues were confederations of scattered cities featuring decentralized decision-making and limited authority over members. In Italy, some cities grew large enough to defend their economic and political interests individually. As a result, they resisted efforts to establish centralized authority under the pope or emperor and instead formed independent city-states. These city-states had external borders similar to territorial states, yet their internal organization was often quite tumultuous with rival factions. The dominant city-state ruled its surrounding villages and lesser cities in an exploitive manner, causing them to continually challenge the authority of the central government.

French cities were much weaker and could not defend themselves as city-states or even in city-leagues. The French kings of the Capetian line were also relatively weak and possessed relatively modest territorial holdings, having granted most of the kingdom as fiefdoms. As a result, the French kings and urban elite banded together against their common threat, the nobles. The cities supported increased central control under the monarchy and, in return, the kings protected the cities against the nobles. The monarchy and urban elite also had a shared interest in fostering favorable economic conditions, including regular systems of taxation, centralized bureaucracy, uniform legal systems, a standing military, transportation infrastructures, and reducing the influence of the nobility and church. By around 1300, cooperation between the French kings and cities fostered the emergence of a territorial state featuring relatively clear external borders and centralized internal sovereignty exercised by a monarch.

These three different forms of political organization co-existed as feudalism faded. Gradually, territorial states proved more effective in providing the security and stability desired by political and economic elites. By 1500, territorial states ruled by hereditary monarchs were coalescing across Europe. Backed by increasingly centralized administrations and improved military capabilities, some monarchs even claimed absolute authority based on a divine right to rule. These absolutist monarchies were a compromise of sorts that aimed at reconciling certain aspects of feudalism, namely monarchical rule and privilege, with the growing economic and political power of an urban-based middle class. The incompatibility of the feudal versus absolutist territorial systems eventually came to a head as the Reformation and wars of religion embroiled much of Europe in over a century of bitter conflict.

The Peace of Augsburg in 1555 and the Peace of Westphalia in 1648 aimed to resolve these conflicts by outlining the basic principles for a new territorial-political order—the modern state system. After decades of widespread conflict, European states agreed to recognize each other as possessing exclusive authority over specific territories. This had three main implications. The first was the notion of territorial sovereignty—or the principle that states would be free to govern their territories without outside interference. Second, states would be regarded as the only institutions that could legitimately engage in international diplomacy and war. Finally, the claim and exercise of territorial sovereignty required that states mark the precise borders of which lands, populations, and resources were included in their territories and which were excluded. Ambiguous frontiers may have been compatible with the flexible and overlapping nature of territorial control in medieval Europe, but they fit poorly in this new context of state sovereignty.

From natural borders to national borders

Supported by burgeoning capitalist economies, expanding government bureaucracies, advances in surveying and cartography, and the emergence of modern nationalism, European states gradually acquired the ability to measure and demarcate the precise limits of their sovereign territories. Yet it was unclear what criteria should determine where those boundaries should be drawn. The idea that borders should follow natural features like rivers or mountain ranges predominated during the seventeenth and eighteenth centuries. The idea of "natural" borders seemed like an objective and logical approach for determining state territories, but in practice individuals tended to focus on whatever natural features supported their geopolitical goals. French writers, for example, often argued that the Rhine River was France's natural eastern limit, which just happened to coincide with French efforts to annex territories along the Rhine. Most other countries likewise focused on natural features beyond their current borders as justifying territorial expansion, while few argued that nature favored borders that shrank their territory. The idea of natural borders commonly served as little more than an excuse for territorial expansion.

Although claiming absolute power, monarchs faced growing demands for greater political participation and representation. Control of the state gradually shifted to democratically elected governments as monarchs were either overthrown or reduced to figureheads. This was an uneven transformation involving many factors, but perhaps the most important was the emergence of nationalism as a mass phenomenon. The rise of modern nationalism transformed interpretations of state sovereignty. If a nation is defined as a group of people who believe they constitute a unique cultural grouping based on shared culture, language, history, and the like, then nationalism is a political ideology that assumes the nation commands the primary allegiance of its members and possesses an intrinsic right to self-determination within its own sovereign state. Previously, the monarch was the embodiment of the state and sovereignty, so much so that the words sovereign and monarch were synonymous. This notion was exemplified by the French king Louis XIV who, shortly before his death, allegedly remarked "I am the State" (*L'état c'est moi*). Following the American and French Revolutions as well as the steady expansion of parliamentary rule in England, this idea of monarchical sovereignty was gradually replaced by popular or national sovereignty. The state came to embody the sovereignty of the nation, not the monarch. The idea of the nation-state—where the political borders of the state would coincide with the cultural boundaries of the nation—became the ideal, although not the norm, by the end of the nineteenth century.

The formation and diffusion of the idea of a German fatherland illustrates this shift from monarchical to national sovereignty. As aspirations for national sovereignty spread in the wake of the French Revolution, ethnic Germans found themselves fragmented between numerous kingdoms, duchies, principalities, and free cities. As a result, German nationalists called for the creation of a unified German nation-state, raising the obvious question of which areas should be included in that new Germany. German writer Ernst Moritz Arndt provided one of the most popular answers, arguing that the German fatherland encompassed "Wherever is heard the German tongue, And German hymns to God are sung!" Rather than relying on natural features to define this proposed state, Arndt and other German nationalists believed ethnolinguistic criteria should be paramount; that is, Germany should include all German-speaking lands. Although a German state was created in 1871, significant German-speaking populations were excluded. Expanding Germany to include all ethnic Germans would remain a major aspiration for nationalist groups, including the Nazi Party, with disastrous results.

As this new idea of state sovereignty gained acceptance, it became increasingly common to argue that borders should follow ethnolinguistic divisions, instead of natural landmarks. The borders of the state of France would thereby include all French populations and lands. Italians also pushed for a new state that unified their nation, while numerous nationalist movements challenged the multinational Austro-Hungarian, Ottoman, and Russian Empires that covered most of Eastern Europe. What began as a state system based on monarchial sovereignty over a royal domain gradually transformed into a state system—or perhaps better described as a nation-state system—based on national sovereignty over an ethnolinguistic homeland.

Nation, state, and nation-state

The idea of drawing borders to achieve national sovereignty would prove difficult to implement, as the peace negotiations following World War I made clear. US President Woodrow Wilson's Fourteen Points for ending the war and preventing future conflict reflected the new view that state borders should correspond to ethnolinguistic homelands. Among his proposals, Wilson called for redrawing the borders of Italy "along clearly recognizable lines of nationality," providing the Turkish people "secure sovereignty," and creating a new Polish state including "the territories inhabited by indisputably Polish populations" with "political and economic independence and territorial integrity." Finally, Wilson proposed an "association of nations . . . for the purpose of affording mutual guarantees of political independence and territorial integrity to great and small states alike."

Wilson's proposals repeated established Westphalian principles combined with the new language of nationalism. States were still regarded as independent entities possessing sovereignty over distinct territories, but their borders should reflect differences in nationality and language. Multinational states, like the Ottoman and Austro-Hungarian Empires, were seen as antiquated. These would be replaced by new nation-states; that is, a Poland would be created for the Poles, a Hungary for the Hungarians, and so forth. Yet reconciling state borders with nationality and language proved just as difficult, subjective, and contentious as attempts to set international borders based on natural features. Ethnolinguistic groups rarely have sharp dividing lines, leading to arguments over which nation had the strongest claim to particular territories. Much of European political history since 1800 involves efforts to revise the region's feudal and absolutist borders to fit this new nation-state framework. Borders played a dual role in these processes as both causes and effects. In some ways, national or ethnic identities provided the rationale for marking new state borders, but in other ways, new state borders had the effect of furthering the creation of new national and ethnic identities. Unfortunately, these bordering processes were marked by massive death, dislocation, and numerous territorial revisions, especially in the aftermath of the World Wars.

As a result, Europe evolved from a region of hereditary monarchies to a region of democratic states whose borders generally correspond with larger nationality groups. Poland during the interwar period, for instance, was a multinational state where approximately 60 percent of the population identified as Polish. The brutality of World War II brought significant demographic and territorial adjustment so that Poland is now 96 percent Polish. The upheavals of the post-Cold War period marked a continuation of this trend as the Soviet Union, Czechoslovakia, and Yugoslavia split apart largely along ethnolinguistic lines. This is not to suggest that Europe's current borders correspond perfectly to national differences or that Europe's borders have been finalized, as evident in the bloody clashes between Ukraine and Russia. Additionally, significant minority populations remain across Europe, and there are several nationalist movements that advocate border changes to create additional nation-states. Prominent cases include the Catalans and Basque in Spain, the Dutch in Belgium, and the Scots in the United Kingdom.

These continuing efforts to adjust borders and reallocate territory highlight an inherent tension within the foundation of our contemporary international system, namely the conflict between guarantees of state sovereignty and territorial integrity on one side and recognition of a national right of self-determination on the other. It is extremely difficult to reconcile the muddled cultural boundaries between ethnolinguistic groups with the precise political borders required by sovereign states. This contradiction is present in the founding charter of the United Nations. Established in 1945, the United Nations was intended to further international cooperation, especially the goal of preventing war. Article 1 of the charter outlines the purposes of the United Nations, including promoting "friendly relations among *nations* based on respect for the principle of equal rights and self-determination of *peoples*" and serving as a "centre for harmonizing the action of *nations* in the attainment of these common ends." But then Article 3 limits membership to "peace-loving *states,*" while Article 2 recognizes the "principle of the sovereign equality" for all member states and prohibits the United Nations from interfering in issues "within the domestic jurisdiction of any *state*" (italics added). So, despite its name, the United Nations is not an organization of nations but rather of states. The states of Turkey and China are recognized as equal, sovereign entities, but the Kurdish, Tibetan, and Uyghur nations are not. This basic contradiction is a primary cause for much contemporary international and civil conflict.

Colonialism and sovereignty

During this transition from feudal to absolutist to territorial nation-states in Europe, Europeans were simultaneously engaged in colonial expansion abroad. Colonialism refers to a process through which a state establishes direct political and economic control over territories beyond its commonly recognized borders. European colonialism was motivated by diverse factors including economic competition, geopolitical rivalries, missionary passions, and settlers seeking greater opportunity. Regardless of the exact motivations, the result was usually an unequal relationship that benefited the colonial power and its settlers more than the colonial territory and its indigenous inhabitants. The establishment of European colonial control over much of Africa, Asia, and the Americas brought dramatic and often destructive changes to colonial lands, societies, and economies, including the imposition of European norms of sovereignty, territoriality, and borders. Although these non-European societies had quite sophisticated territorial practices, it was largely the political and geographical notions championed by Europeans and imposed through colonialism that provided the basis for the modern state system.

The expansion of European forms of territorial organization was integral to colonialism. As one of the first steps in establishing sovereignty over these new territories, European states began mapping and reorganizing those lands to conform to the territorial state model that was emerging back in Europe. From the European perspective, colonial territories were basically "empty" lands to be claimed despite the obvious presence of established populations, societies, economies, and governments. Indigenous approaches to land ownership, resource access, and mobility would be radically transformed or obliterated.

The Treaty of Tordesillas in 1494 was one of the first attempts to impose European notions of territorial sovereignty beyond Europe. Spain and Portugal, the leading colonial powers at the time, hoped to divide all non-European territories between themselves and prevent claims by other European countries. The treaty designated a meridian running roughly through the middle of the northern Atlantic Ocean and South America. Spain claimed all lands to the west of that meridian, which included most of the Americas, while Portuguese claims extended to the east, encompassing Brazil, Africa, and Asia. The Treaty of Zaragoza in 1529 set another meridian on the other side of the world that divided Spanish and Portuguese claims in the Pacific Ocean. Other European powers rejected the treaties, but they nonetheless marked the beginnings of a long-term process in which the European territorial state became the global norm.

Colonialism was an uneven process in terms of the duration, scope, and effectiveness of European control. In general, European colonialism can be divided into two phases. The first phase began in the late fifteenth century. Because Europe's emerging territorial states often lacked the resources to fund large overseas expeditions, many colonial efforts began as semiprivate commercial ventures led by chartered companies that pooled private investment to finance the exploration, conquest, and administration of colonial territories. In return, European governments granted these companies special economic privileges, such as trade monopolies with the colony, to generate profits for the investors. This commercial colonialism was a somewhat indirect form of control where company officials basically functioned as colonial administrators working on behalf of a government back in Europe.

Some of these chartered companies, like the British East India Company, became quite powerful and controlled military forces that rivaled some European states. Nevertheless, company officials controlled relatively modest territories, usually small coastal outposts where they conducted trade with indigenous merchants and rulers. These companies were not primarily interested in controlling territory. Rather, they were interested in controlling the flow of trade goods and people, including slaves, between the colonies and Europe, as well as other regions if profitable. As a result, the companies were largely unconcerned with establishing expansive territorial claims or clear borders. Those were unnecessary expenses, so long as trade goods continued to flow and rival companies kept their distance.

Despite the focus on controlling the movement of goods, chartered companies gradually found themselves administering ever larger territories, often as an ad hoc strategy to block rival claims and maintain their monopolies over local markets and resources. Chartered companies were gradually overwhelmed by the burdens of government. At the same time, states in Europe were increasing their abilities to exercise effective territorial control. This marked the beginnings of a second stage in European colonialism as earlier forms of indirect rule gave way to European states assuming direct and formal control over colonial administration. This transition occurred earlier in the Americas than in Africa and Asia, but European states had established direct sovereignty over most of their colonial territories by 1900.

Creating colonial borders

The shift from commercial to state colonialism greatly expanded the territory controlled by European powers. This transition also entailed the extension of the territorial state model to colonial territories and subsequently the need to clearly demarcate borders. This was perhaps most evident during the so-called Scramble for Africa. After assuming more direct control from the chartered companies during the nineteenth century, European states moved quickly to claim Africa's interior regions. There was a real danger that competition could lead to war, so European leaders met at the Berlin Conference in 1884–85 to partition Africa. It was through this conference and later negotiations that European leaders largely created Africa's modern political borders. They did so with limited information about the lands and peoples they were dividing and without involving Africans. A similar process was underway across much of Asia. Most states were bounded by relatively vague and ambiguous frontiers at the beginning of the sixteenth century, but the situation had changed dramatically by 1900 as colonial powers hurried to mark the exact limits of their territorial claims.

European colonial expansion and border demarcation even impacted areas that were never subject to colonial control. During the nineteenth century, Britain and France were expanding their colonial possessions across Southeast Asia until only the kingdom of Siam—roughly modern-day Thailand—remained independent. The Europeans were operating according to notions of absolute sovereignty, but the Siamese government was structured around a form of shared sovereignty between the central monarch and local rulers in outlying areas. British envoys repeatedly contacted Siamese rulers to negotiate a border between their territories, but the Siamese monarchs had little interest in marking external borders, which mattered little so long as the local rulers remained loyal and served the central government. After continuous European encroachment into their territory, the Siamese government eventually relented and adopted western notions of borders to negotiate clear boundaries for their sovereign territory. This created a "geo-body"—or a territorial outline of a homeland not unlike a logo—that fostered the formation of a Thai state and nation.

Westerners generally saw colonialism as "civilization" advancing and bringing various benefits to colonial subjects, including new ideas of territory, sovereignty, governance, property, and citizenship. This also provided a convenient justification for conquering more land and controlling resources. It also ignored how uneven and incomplete colonial control was in many places. The expansion of European control usually depended on the cooperation or co-optation of some local elites. In British India, sovereignty remained somewhat convoluted even after the British state assumed direct control of colonial administration in 1858. The British government exercised overall sovereignty, but hundreds of Indian princes remained largely autonomous concerning domestic affairs within their territories, such as taxation and law enforcement. As a result, British officials organized several surveys to mark the boundaries between British and princely jurisdictions. Although world maps generally depict colonies with solid colors suggesting political uniformity, arrangements of overlapping or shared sovereignty were quite common.

Colonial rule may have been incomplete and short-lived in some areas, but the imposition of the territorial state model had far-reaching implications. Indeed, most contemporary international borders outside of Europe were imposed through colonialism. Some borders were imposed by victorious states on their defeated rivals. Others originated from agreements between states, such as at the Berlin Conference. In either case, the borders between respective colonial territories, such as between British

4. This map shows European colonial territories in Africa, ca. 1710.

and French colonies, eventually became international borders
between sovereign states, especially during the wave of
decolonization following World War II. Other international
borders emerged when larger colonial territories were partitioned

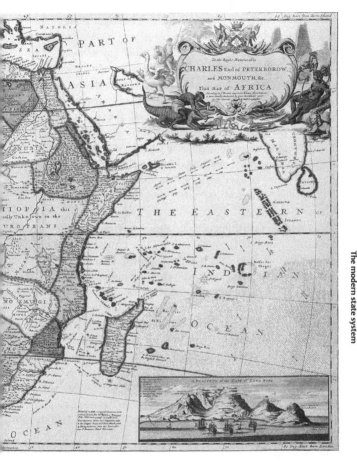

along former internal administrative boundaries, such as in French Indochina or French West Africa.

The British withdrawal from India in 1947 provided one of the most traumatic transitions from colonial rule to independence.

The external borders of British India were largely set, but disagreements between Hindus and Muslims precluded a simple transfer of sovereignty. As a result, the British government decided to partition British India into two states: India, which would include predominantly Hindu areas, and Pakistan, which included two predominantly Muslim areas separated by approximately 900 miles (1,450 kilometers). It was impossible to draw lines perfectly dividing Muslim and Hindu communities into their respective states, so millions of people found themselves on the "wrong" side of the new India/Pakistan border. Many people migrated to the "correct" side, but tragically, thousands were killed in the ensuing upheaval and violence.

The problems of partition in South Asia were further complicated by the semi-sovereign princely states. These princes could choose which state to join. Most Hindu princes ruled majority Hindu populations and joined India, while Muslim princes similarly joined Pakistan. The Kashmir region had a majority Muslim population but a Hindu prince who decided the region would join India. This triggered war between India and Pakistan, resulting in the region's continuing partition between the belligerents. Kashmir's status remains unsettled, and with China claiming a portion of the territory, these border disputes are among the most volatile in the world, as shown by periodic skirmishes between India and Pakistan and most recently between India and China along their so-called "Line of Actual Control." Additionally, East Pakistan broke away from the rest of Pakistan as the independent state of Bangladesh following a bloody civil war in 1971.

Given that the number of sovereign states increased from approximately 70 in 1945 to more than 170 by the 1980s, it might seem surprising that so many colonial-era borders were retained. This again highlights the degree to which the modern state system prioritizes the territorial integrity of states over claims of national self-determination. States have been extremely reluctant to redraw international borders fearing it could ultimately undermine their own territorial integrity. The partition of British India also served as a warning that redrawing borders would likely fuel discord and potentially violence. Most contemporary international borders are relatively recent creations, and quite arbitrary ones at that, but they have assumed a degree of permanence and timelessness. This air of permanence is, however, far less applicable to internal borders.

Domestic borders and boundaries

In addition to their external borders, states also contain numerous boundaries that differentiate domestic spaces. Some of these are official administrative or bureaucratic jurisdictions; others mark unofficial or informal social groupings. Official administrative borders are generally easily identifiable, and most people are aware of living within lower-level jurisdictions. Despite this familiarity, internal administrative borders vary greatly from state to state in terms of their function, impact, shape, purpose, and even basic terminology, including cantons, departments, states, republics, provinces, oblasts, *aimags*, etc. These variations reflect different assumptions concerning the exercise of sovereignty, as well as practical considerations like population density, cultural differences, and natural features.

The form and function of the fifty states in the United States offers an interesting case in point. The formation of the original thirteen colonies reflected varied political and practical considerations of the British monarch when granting colonial charters. As a result, the territories and populations of the colonies differed significantly. In response, the US Constitution ensured some measure of equal representation among the states by creating a senate with two senators per state. The US government later endeavored to create states of roughly equal size based on geometric ideals in its newer western territories. The result is that eastern states feature more irregular shapes and greater variation in territorial size compared with western states.

The United States also adopted a federal system of government, meaning that sovereignty is shared between the central government and states. The US Constitution limits the power of the central government to certain functions like diplomacy and defense, while states retain sovereignty over other functions like education and law enforcement. Many other countries have adopted federal systems granting lower-level governments varying degrees of responsibility, authority, and representation. In contrast to federalism, unitary states feature the concentration of sovereignty in a single governmental unit. In unitary states, lower-level jurisdictions merely implement policies made by the central government, and they have little or no authority to make independent policy decisions.

The desire to achieve equality among the states meant that the United States drew many of its internal political borders as though operating on a blank canvas. In contrast, many administrative borders reflect ethnic or tribal communities, historic realms of feudal lords, estates of princes, or the legacy of colonialism. Even in diminished form, these borders have endured and play powerful roles in shaping political participation, the provision of government services, and cultural identity. This is not to say that borders and the territories they demarcate are permanent—historic atlases provide numerous examples of defunct states and relic borders. The point, however, is that every border has a story. Every line on a map, every marker in the landscape, was derived from some complex negotiation of power, culture, and identity.

The formation of domestic borders and jurisdictions often institutionalizes significant spatial differences in educational opportunities, political representation, government services, or financial services. Unfortunately, the processes creating these borders are highly susceptible to political manipulation. Gerrymandering refers to the drawing of borders with the intent to advantage a specific group. Legislative districts in the United

States are commonly and controversially gerrymandered to make sure that a particular political, racial, or socioeconomic group constitutes a majority and is therefore favored to win any election. Censuses that collect economic, racial, and cultural data often provide the basis for these politically motivated borderings. Comparable situations are easy to find. The division between Israeli and Palestinian territories resembles an archipelago where ethnoreligious segregation has de jure sanction, while several states of the former Soviet Union have recently revised electoral districts to reduce the political influence of their Russian-speaking minority populations.

These geographies of difference extend beyond official demarcations to include various informal socioeconomic and cultural boundaries, which reflect and reinforce social categories of wealth, power, and privilege in both developed and developing countries. Residential patterns in American cities are closely tied to economic and racial differences with lower-income minority groups often concentrated in inner-city neighborhoods. Similar patterns of residential segregation are evident in European cities with lower-income immigrant groups clustered in specific districts. Both cases highlight the complex political, cultural, and socioeconomic forces that create informal spatial restrictions on residential choices for minority and low-income groups in developed countries.

Gated communities further perpetuate residential segregation. As the term suggests, gated communities use walls, fences, checkpoints, guards, and surveillance equipment to create spaces of security and privilege. Initially designed for wealthy households in developed countries, gated communities have become increasingly common in other regions. The proliferation of gated communities reflects in part the growth of high-income households in developing countries, but they often contain high concentrations of foreign nationals. These range from retirement communities in Central America and the Caribbean catering to wealthy retirees from developed countries to residential compounds for expatriate workers in Saudi Arabia and other Gulf states to the stylish condos for the nouveau riche across developing countries. These residential borders help establish and reinforce social differences, as well as setting parameters for daily mobility, access, and interaction.

Similar residential borders can be found in lower-income neighborhoods. Lacking the security infrastructure and formal status of gated communities, gang territories rely on unofficial demarcations of territorial control to establish some measure of privilege for members within that territory. Graffiti and other displays of "colors" may seem like simple markers of lower socioeconomic status, but they can actually represent the territorial claims of particular groups. In a more benign sense, street banners, murals, and commercial signage can also demarcate zones of privilege and belonging for certain social groups. This can take the form of ethnic neighborhoods, religious

5. Signs and other barriers mark the entrance to a gated community in the United Kingdom.

regions, linguistic communities, and LGTBQ2IA+ areas. Numerous "Chinatowns" across North America, Asia, and Europe are often demarcated by arched entranceways and bilingual signs that associate those neighborhoods with a particular ethnic community. These types of ethnic enclaves are common features of cities with large immigrant populations, such as "Little India" in Singapore or "Japantown" in São Paulo. These borders lack official standing and are rarely on official maps, but they are based on the same principles of territoriality as international borders by designating spaces as "belonging" to a specific social group and obliging certain norms, as well as limiting access to others.

The processes of globalization and gentrification suggest a gradual detachment between specific identities and specific regions, threatening long-standing connections between people and place. Yet new international linkages and demographic shifts present as many possibilities for "re-territorialization" as "de-territorialization." Many groups, ranging from ethnic communities to socioeconomic classes, are likely to respond to these broader changes by cordoning themselves off within socioeconomically or culturally homogeneous areas. In this context, internal administrative borders and informal social boundaries provide a means for negotiating new forms of cultural, political, and socioeconomic belonging. The prospect for significant "re-bordering" within states, therefore, coincides with new realities for borders between states.

Chapter 4
The practice of bordering

Border researchers have traditionally focused on the actual geographic location of international borders and their associated signs, fences, walls, checkpoints, and other barriers. More recently, scholarly attention has shifted to understanding why borders exist in the first place and how borders differ from place to place, over time, and for different people. As a result, borders are now commonly understood as processes, rather than things. Instead of just a noun, the word "border" has become the root of a verb— "bordering." That new perspective raises many questions: Do good borders make good neighbors, and if so, what constitutes a good border? How can the demands for state sovereignty versus national self-determination be reconciled? Should the territorial state remain the dominant framework for organizing politics, economics, society, and international relations, as well as addressing pressing issues such as conserving biodiversity, enforcing human rights, and mitigating climate change? Is a borderless world a laudable goal, and if so, what would that entail? These are just a few of the questions raised by the complicated and contradictory nature of borders and bordering in the twenty-first century.

International borders in transition

The dramatic geopolitical changes of the late 1980s and 1990s—most notably the collapse of the Soviet Union, the fall of the Berlin Wall, the advent of the internet, and momentous international economic integration (EU, NAFTA, ASEAN, etc.)— triggered a broad reassessment of borders and their utility, implementation, and meaning among politicians, business leaders, academics, and the general public. Speaking before the US Congress in 1990, US President George H. W. Bush proclaimed that "a new world order" was emerging based on shared values of freedom, justice, and peace, which would foster greater international cooperation. Many disagreed with President Bush's optimism, but there was general consensus that the dawn of the twenty-first century would mark a turning point for borders.

Much of the discussion revolved around the idea of globalization—a catch-all phrase encapsulating a broad range of political, socioeconomic, technological, and environmental processes that facilitate interaction and integration across international borders. Many scholars, politicians, business leaders, and activists believed that globalization would erode the importance of nation-states and their borders. In terms of business, communications, and information technology, some predicted globalization would improve economic efficiency, technological diffusion, and overall standards of living. Others worried about economic decline in developed countries as companies shifted jobs overseas or surges of immigration increased competition for limited economic opportunities and undermined social cohesion.

The terrorist attacks of the early 2000s combined with large-scale migrations of people fleeing conflict zones, escaping environmental disasters, or simply seeking economic opportunities throughout the 2010s fueled populist and isolationist sentiments in many societies. In response, numerous governments sought to strengthen border controls. The United States has constructed new walls along its border with Mexico, while governments stretching across Europe, Africa, and Asia have launched similar initiatives. Border controls are not, however, restricted to walls, fences, and border patrols but are also evident in more robust requirements for inspection, documentation, surveillance, and censorship at airports, ports, public buildings, websites, and many other places far removed from traditional border checkpoints.

These new border controls serve to limit or at least better regulate the movement of people, things, and information, but it is important to recognize that the varied exchanges lumped together as globalization entail contradictory trends encompassing both security and opportunity. Globalization may appear as an unprecedented challenge to the nation-state system, but state borders and territorial sovereignty have always been uneven, elastic, and a bit messy. The shifting balance between security and opportunity was pivotal in the formation of the nation-state system, as well as the advent of colonialism and mercantilist capitalism. Those historical precedents of human connectivity and global integration helped set the stage for globalization, but the intensity of contemporary interaction and interdependency among regions, states, and continents is unique and carries profound implications for notions of territoriality and processes of bordering.

Globalization and territory

The growth of globalization has led many people to question the overriding primacy of state territorial sovereignty. Some even wonder whether nation-states will remain the dominant decision-makers on the international stage or might be gradually replaced by various international, substate, and private actors. The fact that standards of living and life opportunities for most people correspond closely to one's country of birth has given rise to the phrases "birthright lottery" and "global apartheid" to denounce the immorality and randomness of such disparities. In response, some advocate for policies promoting global solidarity, open borders, or even "no borders" justified through varied theoretical perspectives, including cosmopolitanism, humanitarianism, egalitarianism, libertarianism, religion, socialism, and even anarchism.

In contrast, there are proponents for borders and bordering within many of those same schools of thought. These scholars ground their arguments in notions of practicality, responsibility, and the zero-sum framework undergirding territorial sovereignty; that is, a state can only gain territory if another state loses territory. This has been and will almost certainly remain a prime source of conflict. Indeed, much of international law aims at preventing or resolving territorial disputes. Yet it is important to recognize that international law developed within the overall framework of the modern nation-state system. Returning to the premise of the "territorial trap," international law takes the link between state, sovereignty, and territory as a given and unbreakable and, unsurprisingly, almost always supports the status quo regarding international borders.

In fact, the primacy of a state's territorial integrity is explicitly codified by most international organizations. The United Nations charter (Art. 2, Para. 4) holds that "all Members shall refrain in their international relations from the threat or use of force against the territorial integrity or political independence of any state." Similar language is present in the founding charters of the Arab League (1945), the Organization of Islamic Cooperation (1969), and the African Union (2000). Some international organizations, among them NATO and the European Union, condition membership on resolving any external border disputes.

This privileging of territorial integrity ignores the fact that the formation of modern international borders was an uneven and arbitrary process that often divided national groups between multiple states or amalgamated multiple national groups within a single state. This basic disconnect between the cultural borders of nations and the political borders of states has fueled a great deal of violence over the past two centuries and continues to drive many conflicts today. The conflict between Russia and Ukraine stems in large part from the significant ethnic Russian populations that reside in Crimea and eastern Ukraine, while the civil war in Ethiopia's Tigray region pits several different national groups against each other.

Changing state borders to better align with national differences seems an obvious solution. Yet there are at least three major obstacles. First, national groups often have overlapping senses of territoriality, meaning two or more nations claim the same land as their homeland. Second, the emotional fervor of nationalism increases the risk that groups will resort to confrontation, rather than compromise, in settling their claims. Finally, states are reluctant to undermine the sovereignty of other states in fear that their own sovereignty could be undermined in response. This quid pro quo dynamic derives from the fact that state sovereignty is based on mutual recognition. Since the peak of decolonization after World War II, the international community has generally recognized new states or significant border changes only in response to extraordinary events, like the collapse of Yugoslavia and the Soviet Union or decades of violence as in Eritrea and South Sudan.

Despite predictions of a borderless world, much of our daily lives—from human rights and national identity to natural resources and standards of living—remains fundamentally and inescapably linked to territory. Therefore, the prospect of political, cultural, or socioeconomic power completely detaching from territory is unlikely. In fact, states have shown remarkable determination in maintaining and strengthening territorial sovereignty through the first decades of the twenty-first century. Rather than moving in a single direction, borders are being transformed in meaning and function by the contradictory pressures of economic integration, security concerns, and various "foreign" influences. In essence, borders still matter but are taking on new roles and increasingly understood in broader global contexts. Security, nevertheless, remains among the most prominent roles of borders.

Borders and security

Borders provide a powerful symbolic and practical means of dividing "us" and "ours" from "them" and "theirs." The very concept of national security and the state's capacity—some would argue obligation—to use force to defend its interests often centers on border enforcement. Traditional perspectives on national security have, therefore, conferred several roles on borders and borderlands—regions directly adjacent to borders—including serving as sites for gauging the strengths, weaknesses, and intentions of neighboring states, as well as providing buffer zones protecting the state's core. Borders and borderlands may, therefore, function as symbols of socioeconomic stability, as well as potential flashpoints for problems ranging from cultural change to economic downturns to any number of other social ills. In short, the border—those living near it, those crossing it, and its popular perception—help shape political attitudes and policies. This point is worth elaborating as the concept of security is increasingly understood in much broader contexts.

Border security remains a means to prevent foreign military incursions, but today it is more likely to be discussed in relation to countering large-scale and undocumented immigration and crime. Article 13 of the Declaration of Human Rights proclaims that "everyone lawfully within the territory of a State shall, within that territory, have the right to liberty of movement and freedom to choose their residence." Moreover, "everyone shall be free to leave any country, including their own." This assertion of freedom of movement, even across borders, is qualified by the subsequent declaration that "the above-mentioned rights shall not be subject to any restrictions except those which are provided by law, are necessary to protect national security, public order, public health or morals or the rights and freedoms of others, and are consistent with the other rights recognized in the present Covenant." Through this qualification, the state determines the criteria of national security, public order, and morals in setting border policies for migration and travel.

The expansion of border security to encompass a broader range of perceived threats has led to a corresponding broader definition of "border space" and border control in the twenty-first century. For instance, Australia has increased maritime patrols to interdict potential immigrants and asylum seekers far beyond Australian shores and place them in detention centers. The EU's Frontex program has pursued a similar approach to reducing the number of migrants heading to Europe by partnering with governments in the Sahel to enforce European immigration policies, essentially by deterring immigrants at the source. In early 2021 US Vice President Kamala Harris traveled to Central America to discourage would-be migrants from attempting to enter the United States illegally. The fact-finding mission also sought to ascertain how the United States could work with Central American governments to reduce corruption, enhance economic opportunity, and thereby keep Central Americans in place. Ironically, these efforts to partner with politically fragile governments to restrict mobility also risks destabilizing those states and inadvertently fueling migration.

The above cases highlight how governments increasingly regard border policies as tools to address concerns across their entire territories and not just the specific interests of borderlands. These more systematic and comprehensive approaches to border security increasingly involve coordination with neighboring and even distant states. Toward this end, new perceptions of borders as spaces, instead of simple lines on a map, focus attention on the array of new conditions, institutions, and actors influencing how easy it is to cross a border. In essence, borders function as filters with a variety of different gateways and settings. Some of those operate at the physical border between two states, others play out at remote sites of bureaucratic authority, such as embassies, consulates, and transit portals in airports, and still others happen by proxy through other governments.

As human creations varying across space and time, borders and borderlands are intimately linked to legal, governmental, historical, political, and socioeconomic contexts. Specific threats and pressures may increase or decrease the level of border enforcement for a given border at a given time, and sometimes enforcement may even differ at various points along the same border. The US-Mexico border provides one interesting case in point. The presence of violent drug cartels in Ciudad Juarez, Mexico, triggered increased monitoring and restrictions on movement into and out of El Paso, Texas. In contrast, the border between Tijuana, Mexico, and San Diego, California, about 620 miles (1,000 km) to the west, remains relatively open for tourists and commerce. The relativity of border enforcement is made equally clear by the ease with which economic elites cross borders in contrast to the less wealthy. This makes clear that borders function as imperfect filters reflecting both inclusive and exclusive policies, which differentiate between types of people, materials, motives, and information. Rather than being a concept inherent to the pursuit of freer movement, "border permeability" is a variable category that fluctuates between relative closure and relative openness.

Amid increasingly intense cross-border flows, the goal must be to construct "good" borders. Toward this end, many would argue that "good" borders generally feature open communication, formal demarcation agreements, standing boundary commissions, accessible transportation links, and a minimal military or police presence. But "good" borders may also hinder harmful crossings and thereby reduce antagonisms. In those cases, a border's ability to impede flows may be especially beneficial. States are strengthening and extending some border enforcement in the name of security and simultaneously weakening and relaxing other border policies in the name of opportunity as they seek a balance between the threats of terrorism and varied forms of trafficking and the benefits of global economic integration and labor migration.

Contingent sovereignty

The notion of contingent sovereignty illustrates the evolving nature of the nation-state system. Among border scholars, contingent sovereignty refers to the idea that territorial sovereignty—traditionally equated with the inviolable authority of the state—is being challenged by numerous groups arguing that violations of human rights or the proliferation of weapons of mass destruction compel international action. For proponents of contingent sovereignty, states in serious violation of global norms forfeit their territorial sovereignty. The international community is then empowered to disregard that state's sovereignty to enforce various conventions and treaties. The offending state may then be subject to various forms of monitoring and oversight by the international community in order to regain its territorial sovereignty through "good behavior." Specific cases include the Road Map for Peace in the Middle East, the Good Friday Agreement in Northern Ireland, the Naivasha Agreement for Sudan, the Baker Peace Accords for Bosnia, and UN Security Council Resolution 1272 for East Timor.

These new interpretations of state sovereignty and calls for global enforcement of human rights are, nevertheless, a far cry from a patently new system of territorial organization. Would China or the United States ever agree to applying that standard to themselves? Contingent sovereignty and its potential to supersede or suspend territorial sovereignty and international borders are, therefore, generally only applied to weaker states. The problematic nature of contingent sovereignty was laid bare by the international response to the Arab Spring revolts in 2011. Following a UN resolution, NATO members launched air strikes supporting rebels seeking to overthrow the Libyan dictator Muammar Gaddafi. Although technically a violation of Libyan sovereignty since Libya was a UN member, NATO leaders justified the strikes on the grounds that Gaddafi loyalists were engaging in human rights abuses. Therefore, the international community could intervene on the basis of a "responsibility to protect" civilians. The NATO strikes played a critical role in overthrowing Gaddafi's regime.

In contrast, the international community was more reluctant to intervene against the brutal suppression of protesters ordered by Syrian dictator Bashar al-Assad. As the international community finally took sides, the conflict turned into a series of overlapping proxy wars. The United States, France, Saudi Arabia, Qatar, Turkey, and the United Kingdom, along with Syrian rebel forces, pitted themselves against al-Assad's regime. The Lebanese militant group Hezbollah and the Syrian-based Palestinian group PFLP-GC, both supported by Russia and Iran, allied with al-Assad. The situation was further complicated by the fact that significant portions of Syrian territory were controlled by the Islamic State of Iraq and the Levant (ISIS) between 2014 and 2017. As an internationally recognized terrorist organization, the United States, Russia, France, and the United Kingdom launched military strikes against ISIS. At the time of writing, the Assad regime has re-established control over the country, except for some small pockets in the Idlib region. Western intervention on behalf of human rights was largely ineffectual, and many Syrians continue to face a protracted humanitarian crisis.

State-sponsored repression and violence against civilian populations continues to fuel calls for new interpretations of sovereignty, but violent authoritarian backsliding by governments in the wake of the Arab Spring revolts, Yemen's near collapse as a functioning state, and reluctance to intervene in Ethiopia's conflict with rebels in the Tigray region bode poorly for future international action based on human rights. The doctrine of "responsibility to protect" seems unlikely to rouse meaningful action given that American efforts to intervene in Somalia in the early 1990s did little to alleviate civilian suffering and the twenty-year American-led war in Afghanistan ended in an abrupt and humiliating withdrawal and the restoration of a repressive Taliban government in 2021. Direct foreign intervention is probably unlikely unless the offending regime is very weak—lacking formidable military forces or powerful allies—but then again, the Taliban lacked formidable forces and allies. When faced with a powerful aggressor, such as Russia's bloody invasion of Ukraine in 2022, foreign intervention is more likely to take indirect forms, such as military and humanitarian supplies and economic sanctions.

Interventions to impede the proliferation of nuclear, biological, and chemical weapons are more likely. Authoritarian regimes in Iran and North Korea have clear track records of human rights abuses, but it is mostly their nuclear weapons programs that garner international concern. Israel has undertaken a series of covert actions against Iran's nuclear program and has trained for possible airstrikes should intelligence sources suggest substantial progress is being made. North Korea's efforts to develop a nuclear arsenal have compelled world leaders to broker deals with mercurial president Kim Jong Un, while US President Joseph Biden seeks to reinstate the "Iran deal" reneged by his predecessor. It seems clear that calculations of cost in terms of "blood and treasure" will deter interventionism even if justified by clear violations of international norms and agreements. Similar calculations have long applied to borders within states as well. However, recent public attention to ideals of justice and human dignity have directed long-overdue scrutiny to borderings of minority and indigenous territories.

Indigenous sovereignty and minority territories

The nation-state system was built largely on the notion of *uti possedetis*—as you possess, so you may possess—and western legal traditions of property and ownership—possession being nine-tenths of the law. The concept of "indigenous sovereignty" challenges those notions and the racist doctrines that justified the historical, political, and legal treatment of indigenous and minority land claims. As a result, activists and scholars have called for a broad reassessment of the basic notions of territorial state sovereignty and private property and further noted that premodern notions of overlapping polities, nonhierarchical power structures, and frontiers have no less standing than western ideals of territory, authority, and boundary. From this perspective, the process of state formation imposed severe injustices on indigenous and minority populations that are rarely acknowledged, much less rectified.

For advocates of indigenous sovereignty, the unacknowledged negative impact of border formation on contemporary domestic politics constitutes a moral quandary. Put simply, most states and their territories were demarcated under highly undemocratic circumstances. The formation of state borders generally resulted from unequal power relationships that both reflected and crossed various social boundaries. Even borders demarcated to facilitate the formation of democratic states and civic nations were rarely the product of democratic processes. Ironically, democratic politics were expected to emerge from democratic institutions tied to modern states even though democratic institutions are generally assumed to require a geographically bounded electorate. Within the context of a maturing democratic system, this original imposition of power by one group over another may perpetuate lasting political and socioeconomic inequalities, which in turn may fuel international conflicts, ethnic resentments, and social injustices. In this sense, borders represent the "scars of history"— not only physically across the landscape but symbolically and metaphorically in the minds of various populations.

Many different forms of minority territory resulted from the varying cultural, economic, and political conditions that accompanied the process of state formation. Enclaves and exclaves are good examples. Enclaves are a portion of one state's territory completely surrounded by the territory of another state. In contrast, an exclave is a territory belonging to one state that is not contiguous with the rest of that state. Most exclaves are also enclaves inside another state. Exceptions to these definitions include exclaves that are not completely surrounded by another state, such as the Spanish exclave of Ceuta on the North African coast, or enclaves that do not belong to another state, as is the case for San Marino inside Italy.

Enclaves and exclaves are quintessentially the "scars of history," often generating intense emotion on both sides of the border. The violence in southern Kyrgyzstan in 2010 involved such simmering tensions. However, territorial structure was not the primary cause of the violence. Rather, political alliances, governmental instability, economic hardship, and geopolitical leveraging by neighboring states provided the underlying causes for the clashes between Kyrgyz factions and Uzbeks within Uzbekistani enclaves. That being said, one cannot deny the oddity of these enclave and exclave borders.

The Uzbek enclaves inside Kyrgyzstan, like many borders in the former Soviet Union, stem from the idiosyncratic efforts of dictator Joseph Stalin to create a socialist union from the multinational czarist empire. Fifteen larger ethnic groups, such as the Russians, Ukrainians, and Uzbeks, were given republics within the Soviet Union, while smaller ethnic groups, including the Abkhazians, Chechens, Karakalpaks, Tartars, and dozens of others, received autonomous regions within another ethnic group's republic. These borders were largely symbolic and relatively unproblematic, since the Soviet Union was highly centralized. This situation changed following the collapse of the Soviet Union as republic borders became international borders and autonomous regions gained greater political significance. The bitter clashes between Chechens and Russians during the 1990s and early 2000s, Georgia's conflicts with South Ossetia and

Abkhazia, and recurring fighting between Armenians and Azerbaijanis over Nagorno-Karabakh vividly illustrate the unintended and tragic consequences of such geopolitical contrivances. Today, the Chinese Communist Party subjects the Tibetan and Uyghur autonomous regions to intense policing, surveillance, and oppression to prevent similar secessionist movements from gaining traction.

Autonomous ethnic regions are not, however, unique to socialist states. The United States has a reservation system for much of its indigenous population. There are some 310 reservations across twenty-six different American states, each possessing special jurisdictional rights. Gambling is one of the most visible results of those jurisdictional rights. Until the 1980s, gambling was legal only in Nevada, Atlantic City in New Jersey, and on river boats, but Native Americans have since seized on their special tribal sovereignty to build casinos. Revenues from luring "tourists" have helped counter inordinately low standards of living on reservations, though intense debate exists as to the distribution of gambling proceeds. Some reservations also employ indigenous jurisdictional rights to permit certain cultural practices relating to education, hunting, or the use of substances, such as peyote, that are regulated or illegal under federal law. Moreover, minor crimes committed on the reservations are addressed by tribal councils or courts with limited sentencing capacity, while major crimes are investigated by federal law enforcement officials and adjudicated in federal courts.

Perhaps the most substantive territorial policies relating to indigenous rights are the 1999 Nunavut Act and the Nunavut Land Claims Agreements, which granted autonomy to much of Canada's northernmost territories as Inuit lands. In a lesser but also significant decision, the US Supreme Court in 2020 held that much of the city of Tulsa and eastern Oklahoma had been a reservation of the Muscogee Nation. This decision has the potential to be a watershed in upholding tribal boundaries and treaty obligations ignored or broken throughout US history. Such efforts speak to the power of "land acknowledgment statements" and the relationships between peoples and traditional homelands. Bordering these territories is, thus, a complex process of both inclusion and exclusion, protection and distancing.

Graduated and detached sovereignties

The growing significance of transnational corporations, organizations, and practices has compelled some states to create new domestic borders that facilitate global, neoliberal economic exchange. As a result, states seem to be ceding sovereignty over some segments of the economy, society, and territory to supranational, global, and private entities. This shift constitutes a dramatic restructuring of the relationship between borders, territory, society, and government.

Such a restructuring is evident in the idea of graduated sovereignty, which refers to a state's differential treatment of segments of its population or territory. Special economic zones (SEZs) are examples of states voluntarily limiting their authority over specific spaces. Also called free trade zones, export processing zones, and free economic zones, SEZs are areas subject to more liberal or relaxed economic regulations in hopes of attracting foreign investment and spurring growth. These de facto enclaves range from technical innovations zones to industrial parks to tourist districts. Despite these differences, they all complicate notions of absolute territorial sovereignty and blur the distinction between international and domestic borders. SEZs can be found in virtually every corner of the developing world. India has approved some 378 SEZs with 265 already fully operational by 2020, while China has 14 open coastal cities, 5 special economic cities, and 1 special economic province.

The unique borderings resulting from these recalibrations of sovereignty are also present in western states. State and local governments in the United States have tried to attract foreign businesses like BMW, Honda, Toyota, and Foxconn by offering significant tax reductions and other incentives. Research into these incentive programs has shown that many companies often pay considerably lower land, building, and equipment taxes than established local businesses while the promised economic development has proven more mixed.

Malaysia has followed a similar approach to stimulate economic growth. In addition to granting special privileges to entrepreneurs and foreign investors, the Malaysian government also limited the ability of laborers to unionize and allowed bonded labor in export-orientated industries. This created a multi-tiered system exerting stricter control over manual laborers but minimal regulation over businesses and investors. Such efforts to condition or even abrogate state sovereignty in exchange for economic benefit suggest growing detachment between territory, power, and authority.

Extra-territorial or detached expressions of sovereignty entail control of places beyond the recognized limits of state territorial jurisdictions and provide additional cases of exceptional borderings. Detached sovereignties include things like transit portals in airports, embassies, and ships at sea. Although detached from their state, each functions as bounded sovereign territory of that state to greater and lesser degrees. Immigrant and refugee centers, as well as special detention facilities, also constitute extra-territorial sites of state authority and further challenge the territorial basis of law enforcement and justice.

The US naval base at Guantanamo Bay, Cuba (GITMO), is undoubtedly the best known of these extra-territorial sites. The expansion of the United States in the late nineteenth and early twentieth centuries created questions regarding the territorial scope of the US Constitution. Initially limited to the "United States proper," a 1901 Supreme Court ruling extended a limited number of constitutional rights to citizens living in overseas territories that were controlled by the United States but not intended to become states. During World War II, full constitutional rights were extended to American citizens in territory controlled by the United States but not part of the country proper, such as overseas military bases and the Panama Canal Zone. Foreign nationals occupying those territories were not covered, however.

This provided the basis for the US President George W. Bush's assertion that GITMO was essentially a "rightless" zone for foreign nationals, since it is foreign territory "leased" from Cuba. Therefore, rights to due process, a fair trial, and habeas corpus, as well as the Geneva Convention, did not apply. The base seemed ideally suited to serve as a detention center for suspected "enemy combatants" in the war on terror. The US Supreme Court rejected that argument twice, and in 2009, US President Barack Obama ordered the prison closed and detainees transferred to a correctional facility in Illinois. In 2011, however, Congress countermanded the order by prohibiting the transfer of GITMO prisoners to the mainland or other foreign countries. This effectively forced the detention facility to remain open, although US President Joseph Biden announced in 2021 his intention to close the camp.

Controversy surrounds GITMO's status, but the idea of extra-territorial jurisdiction can have positive connotations as well, such as the power of the International Criminal Court to prosecute people for war crimes, genocide, and other crimes against humanity. Yet the court's ability to prosecute alleged international criminals remains contingent upon the cooperation of sovereign states. As a result, accused criminals can find sanctuary in sympathetic states, such as Saudi Arabia providing sanctuary to Tunisian dictator Zine el-Abidine Ben Ali. The arrest of dictators such as Chile's Augusto Pinochet and Liberia's Charles Taylor represent positive acts of the extra-territorial justice, but the related practice of "rendition"—or the abduction and transfer of people to states practicing enhanced interrogation or "black sites"—poses serious ethical questions and potential for abuse.

Emerging borderlands

Conventionally depicted on maps as solid black lines, borders convey the impression of finality and permanence. Yet significant portions of the approximately 308 land borders separating roughly 190 states are not marked and, in some cases, not even surveyed. Vast new borderlands emerge if air, water, and underground realms are also considered as states and other actors negotiate sovereign control over these new "territories." Out of a total of 417 maritime borders, some 228 borders remain in question. The fact that only 45 percent of potential water borders have been firmly established reflects centuries of debate concerning the extent to which states may claim exclusive sovereignty over the seas.

A state's territorial waters were initially set in the seventeenth century as three miles from shore, which reflected the range of a contemporary cannon. This rather loose standard regarded all waters beyond that three-mile zone as international waters subject to the principle of *mare liberum*—or "freedom of the seas." Being outside the sovereign control of any state, the high seas remained among the last frontiers in the modern world. Nevertheless, in the early twentieth century, state interests in mineral resources, fish stocks, and pollution control prompted various redefinitions of the limits of territorial waters. US President Harry Truman's 1945 declaration of the continental shelf as the natural extension of national territory set a precedent for later seabed resource exploitation.

Other states opted for a twelve-mile zone of maritime sovereignty, while still others have claimed "exclusive economic zones" (EEZs) stretching as far as two hundred nautical miles from the coast. In effect since 1994, the United Nations Convention on the Law of

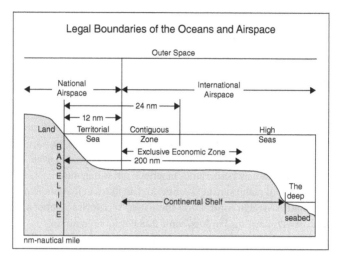

6. The Law of the Sea helps define maritime and airspace borders, as well as seabed rights.

the Sea (UNCLOS) introduced provisions for maritime sovereignty over EEZs and continental shelves, navigation and transit procedures, deep seabed mining, marine environment protection, exploration and research, and dispute resolution. UNCLOS has come under increasing pressure, especially as states seek to exploit oil and natural gas reserves beneath the seabed.

The competing maritime claims in the South China Sea demonstrate how conflicting maritime borders have escalated as international issues. In addition to some of the world's busiest shipping routes, the South China Sea also supports very productive fisheries. Exploration for oil and natural gas has also suggested the possibility of major undersea deposits and raised the commercial and strategic stakes considerably. As a result, there have been heated words and periodic naval skirmishes between rival claimant states, especially between China and its southern neighbors. The matter is further complicated by the presence of Indian and American naval, commercial, and scientific vessels, among other countries. China's island-building projects in the region—accomplished by building out intermittently submerged shoals to support "permanent" bases—provide a foundation for expansive claims to "blue territory." The sheer volume of space claimed by China through these efforts would constitute one of the largest "land" grabs in history.

Another maritime controversy has developed in the Arctic region as the steady retreat of ice pack opens the region for exploration

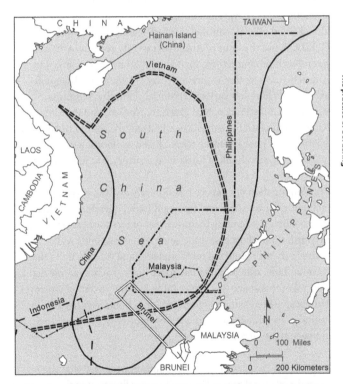

7. **Tensions between littoral states of the South China Sea often relate to overlapping claims to territorial waters.**

and commerce, thereby transforming its strategic importance. In addition to providing shorter shipping routes between many locations in the Northern Hemisphere, an increasingly ice-free Arctic would undoubtedly become a major commercial fishery, and initial exploration indicates significant undersea oil and natural gas deposits. This issue gained international prominence in 2007 when a Russian expedition used a submersible to plant a Russian flag on the seabed at the North Pole to claim the area and its resources as an extension of Russian sovereign territory. Other states bordering the Arctic region, and even some rather distant like China, have launched expeditions to survey the seafloor and strengthen their naval capabilities in the region. Some have already filed official "continental shelf" claims based on the UNCLOS, but the issue of Arctic sovereignty remains contested. For now, these competitive maneuverings have not emerged in Antarctica, where the Antarctic Treaty of 1961 and subsequent agreements have left various territorial claims by Argentina, Australia, Chile, France, New Zealand, Norway, and the United Kingdom in place but unresolved. Instead, the continent has been relatively open for scientific research and exploration, and even hosts the largest chunk of unclaimed territory, Marie Byrd Land.

The demarcation of airspace has similarly evolved with technological advances and raises profound questions concerning the viability of extending the principles of territorial sovereignty skyward. Control over airspace generally corresponds with sovereignty over land and territorial seas and extends up to the Kármán Line, which marks the upper limit of aerodynamic flight. Control of this space was irrelevant before the advent of air travel but became increasingly significant for national security as military aircraft advanced and states vigorously protected their airspace rights accordingly.

In 1955, US President Dwight Eisenhower proposed that the United States and Soviet Union allow overflights of their respective territories for intelligence gathering as a means of assuring that the other was not preparing to attack. Though rejected by the Soviets, the idea was reintroduced by US President George H. W. Bush as a means of building trust between NATO and Warsaw Pact countries. A treaty signed in 1992 entered into force in 2002 and allowed hundreds of missions by various signatories. In 2016, however, limitations began to be imposed in varied circumstances, including Turkey's restriction of a Russian flight over its territory bordering Syria and NATO bases. Russia later restricted flights over portions of Georgia that it controls and has since been accused by the United States of further treaty violations, ultimately leading both Russia and the United States to withdraw from the Open Skies Treaty.

In conjunction with the 2023 controversy pertaining to a Chinese balloon over US airspace, these cases suggest a heightening of airspace borders between states, but they do not necessarily counter the deregulation of air travel and the promotion of the International Civil Aviation Organization (ICAO) initiated in the 1960s. Those efforts made states less assertive in protecting their national airlines and airspace for commercial travel. ICAO facilitates vital cross-border coordination, especially for airports located near borders; however, ICAO serves only as an arbitrator between states. States retain sovereignty over the air above their territories but have generally held to regulations respecting the sanctity of international flights traversing their airspace. The 2021 forced landing of a Ryanair flight at a Minsk airport and subsequent arrest of the journalist Roman Protasevich and student/activist Sofia Sapega by Belarusian security services was a stark violation of international air-travel norms.

Airspace sovereignty, therefore, retains geopolitical significance and is further enhanced by missile defense systems, radar monitoring, and satellite technology. Governments have regularly used the enforcement and violation of airspace as a means of sending messages. Russian incursions into Lithuanian airspace regularly signal Russian sovereignty over the Kaliningrad exclave.

France's refusal to allow American planes to cross its airspace en route to the 1986 bombing of Libya revealed French disapproval of the operation. During 2021, China made multiple daily incursions into Taiwan's airspace, prompting further "saber rattling" by both sides and their allies, while in 2022, most European and North American governments closed their airspaces to Russian aircraft, including commercial airlines.

Popularized as the final frontier, outer space gained clear geopolitical significance following the launch of the first satellite into orbit by the Soviet Union in 1957. This led to the Outer Space Treaty of 1967 between the United States, the Soviet Union, and the United Kingdom, which declared outer space, the moon, and other celestial objects to be "the province of all mankind" not subject to claims of state sovereignty. Now with 112 signatories, the treaty prohibits states from placing nuclear and other weapons of mass destruction in space. The construction of military bases and other military activities on the moon or other celestial bodies is also specifically forbidden. Notably, the treaty does not prohibit the deployment or use of conventional weapons in orbit around the Earth or moon, or just floating in space. This paved the way for the US Strategic Defense Initiative—or "Star Wars" program—in the 1980s that caused great concern that de facto American sovereignty over outer space was imminent. Star Wars was never implemented, but the United States, Russia, and China all currently possess Earth-based weapons systems capable of destroying an opponent's satellites should the need arise and are experimenting with hypersonic weapon systems that greatly surpass the performance of ballistic missile systems.

Space itself is not subject to claims of state sovereignty, but the Outer Space Treaty vested satellites and other spacecraft with state sovereignty similar to ships at sea and embassies in foreign countries. Moreover, the prospect of the commercial exploitation of space further complicates what had previously been the domain of states. Advances in satellite and remote sensing technologies also have a variety of cross-border commercial and cultural applications ranging from rapid dissemination of news via global media conglomerates to the availability of clear aerial images of much of the world's surface via the internet. Satellites have obvious strategic applications for cross-border surveillance. With the erosion of the Open Skies Treaty, intelligence-gathering flights below the Kármán Line would violate state airspace, but satellites orbit above the line, so detailed aerial images can be obtained without violating sovereignty.

Satellites have not been weaponized so far, but they still have direct military implications. The US Global Positioning System (GPS) relies on a network of satellites in geosynchronous orbit to provide location information for planes and ships, as well as to an increasing number of drivers and hikers. GPS also affords a significant advantage for the American military by providing extremely accurate targeting information for various guided munitions. China's BeiDou Navigation Satellite System went operational in 2020, the EU's Galileo system began working in 2016, Russia's GLONASS-K2 system affords global coverage, and India's NavIC provides regional coverage across much of the greater Middle East, East Africa, Asia, and the Indian Ocean. Many military strategists predict that the first shot of the next great war will occur in outer space to blind the opponent. These examples from air, sea, and space highlight how approaches to territory, sovereignty, and borders continue to evolve in response to new technologies, commercial possibilities, and geopolitical concerns.

Chapter 5
Border crossers and border crossings

In the twenty-first century, people are crossing borders with unprecedented frequency and volume. Yet the processes and experiences of crossing borders are quite varied. Some borders are marked only by signs or require a brief document check. Other borders are marked by imposing obstacles or require thorough inspections. Those variations create distinct transitional zones or borderlands, where cultural exchange and hybridization are regarded as mutually beneficial and celebrated in some circumstances but generate anxiety, mistrust, and even hostility in other contexts.

Migrants and refugees

Many contemporary social theorists suggest that the proliferating processes of border-crossing are remaking various categories of social belonging, especially the notion of citizenship. This discussion contrasts the negative connotations commonly associated with words such as immigrants, deportees, and refugees with more positive words such as cosmopolitans, jet-setters, and global citizens. The former categories tend to elicit anxieties over welfare dependency, crime, and socioeconomic competition, while the latter categories suggest free, confident, and enterprising individuals eager to embrace new spaces of interaction, connectivity, and belonging. Those nuanced perspectives complicate commonly accepted understandings of citizenship and identity by demonstrating the shifting nature of categories such as insider and outsider and processes of inclusion and exclusion. Nevertheless, such dichotomies remain powerful and are increasingly evident as the individual human body functions as a site of border enforcement.

Refugees provide a clear example. The United Nations defines refugees as those crossing borders because of a "well-founded fear of being persecuted for reasons of race, religion, nationality, membership of a particular social group, or political opinion." Refugee status is, therefore, explicitly conditional on being outside the country of citizenship and unable or unwilling to rely on protection from that country's legal system. As a result, refugees are entitled to asylum—or the right to remain in a foreign state without the possibility of extradition to their home state. That distinction highlights the harsh reality of differential treatment for those crossing state borders. Those fleeing the oppression of communist or religious dictators differ in migrant categorization from those desperate to escape poverty or environmental degradation.

The role of the state of origin is also evident in the practice of diplomatic immunity, which functions as a form of mobile sovereignty that largely exempts those possessing it from the legal jurisdiction of the state in which they physically reside. These individuals literally and figuratively transplant the sovereignty of their home country into the jurisdiction of the host state. They are "in" another country but not "of" that country and, as such, are largely exempted from the law enforcement of the host society. Tourist, work, and education visas offer similar manifestations of achieving mobile citizenship and national identity by framing the body as sovereign space.

In contrast to a diplomat's immunity or a tourist's official visa, the unauthorized immigrant's body constitutes a violation of sovereignty and the very site of border enforcement through detention or expulsion. The difficulties facing unauthorized immigrants stand in stark contrast to the relative ease with which other people cross boundaries. Those with specialized skills, for instance, often find borders relatively easy to cross, while unskilled laborers generally face greater obstacles. For those concerned with migrant rights, borders constitute troublesome sites of vulnerability. Unauthorized immigrants face the most extreme risks, but all border crossings provide potential opportunities for extortion, exploitation, and abuse. For some scholars and human rights activists, rigorous enforcement of stringent immigration policies violates human dignity and is therefore unethical. Given the costs of welfare, healthcare, law enforcement, education, and other social services, most governments would counter that limiting access to state benefits is necessary for a functioning society.

Arguments opposing migration are inevitably entangled with concerns over social cohesion and solidarity, if not cultural homogeneity. Whether discussing the European Migrant Crisis of 2015–16, the ongoing flow of people from the Caribbean and Latin America across the US southern border, or the three million, with plausible estimates as high as six million, unauthorized migrants in Russia, opponents of immigration often claim that "foreigners" will contribute to an increase in crime and simultaneously erode national traditions. Immigration restrictions have also generated intense controversy in Australia, Canada, Japan, and across Europe. Efforts to restrict African emigration into the European Union involve the construction of new fortifications along Spain's southern borders, as well as increased patrols in the waters separating Europe from Africa. These concrete actions are coupled with a symbolic contrasting of Spanish/European identities versus Moroccan/African identities. Such efforts reveal how borders are far more than lines on the map or locations in a landscape but constitute mobile and contingent processes of exclusion, alienation, and differentiation.

The categorical distinction between "welcome" and "unwelcome" is often shaped by characteristics of the migrant and the destination state. This is not strictly an issue of wealthy states seeking to preserve privilege. What may be termed South-South migration or the movement of people between developing countries—whether avoiding environmental problems, fleeing conflict or disease, or seeking economic opportunity—produces similar concerns of social cohesion and resource scarcity. Different states deal with migrants very differently, as evidenced by the general accommodation and support provided for refugees in Sweden as opposed to the wall building and detention facilities erected in Hungary. In late 2021, Belarusian authorities were widely seen as orchestrating a crisis by channeling flows of immigrants from the Middle East toward the neighboring EU member states of Latvia, Lithuania, and Poland, which enhanced border controls in response. In contrast, Poland and other EU members bordering Ukraine have been relatively accommodating to those fleeing Russia's invasion of that country in 2022. Social

8. **Millions of Ukrainian refugees fled to the EU after Russia's invasion in 2022.**

and legal distinctions between citizen and non-citizen, immigrant and refugee, and voluntary and involuntary migrant constitute geographical categories of belonging, between those "in place" and those "out of place." Borders play central roles in institutionalizing these differences, yet identities are never completely contained by borders.

Transborder identities and communities

Migration studies have traditionally emphasized the one-way and permanent movements of people to new places, such as the migration of Europeans to the Americas. Today, however, patterns of cyclical, return, and seasonal migration between states have become increasingly common. Such migrations inherently involve multiple border-crossings over the course of a lifetime or even within a single year. Coupled with advances in communication and transportation technologies, these circular migration patterns enable the formation of increasingly dense networks across great distances with profound cultural, political, and socioeconomic implications that blur the distinction between "sending" and "receiving" societies.

The term "transnational social fields" refers to organized efforts to facilitate collective action beyond the traditional nation-state system. The idea of hyphenated identity that combines an ethnic identity with current state citizenship, as in Irish American or Korean Kazakhstani, is quite commonplace today. In most instances, such constructions are banal and merely assert a sense of shared heritage. In other instances, the concept of sovereignty is profoundly tested as diasporas—usually defined as ethnic groups living outside their historic national homelands with the desire to someday return—are increasingly able to influence domestic debates in their homelands and conversely elites residing in the homeland are better able to mobilize diasporas. The increasingly robust nature of contemporary transnational social fields draws the very definition of diaspora into question as many dispersed groups show minimal interest in returning to their historic homelands. Instead, they adopt a hybridized identity that constitutes a status of national belonging that is "both/and" rather than "either/or" in relation to their group's historic place of origin and the individual's current place of residence.

Some of these transnational social fields play prominent roles bridging sending and receiving societies, such as Algerian populations in France, Turks in Germany, and both Armenian and Chinese diasporic communities in many states. Others have a lower profile, such as growing Asian communities in Australia, Lebanese merchant groups in Africa, and Tajik labor migrants in Russia, among many others. These diverse diasporic communities operate within and create radically different socioeconomic, political, and cultural networks. In each case, political influence, cultural exchange, and economic remittances have affected both the sending and receiving societies, at times in very profound ways, by creating dependencies and vesting power in new places and peoples.

The growing influence of transnational social fields has been especially evident across Central Eurasia since the collapse of the Soviet Union. Prominent among these diasporic networks are Armenian elites, both domestically and abroad. Some Armenian émigrés were highly nationalized and wished to return to the newly independent Armenian state formed when the Soviet Union dissolved. Others in this diasporic community were committed to their communities abroad. As a result, diasporic organizations, such as the Armenian Pan-National Movement, were established to manage the flow of remittances, form development programs, and coordinate educational and cultural exchanges. The success of those organizations led other newly independent nations across Central Eurasia to establish their own diasporic outreach associations, each harnessing some level of national attachment across their dispersed communities, including the World Azerbaijani Congress, World Association of

Kazakhs, and the Crimean Tatar National Assembly, to name a few examples.

As in other developing regions, such cross-border linkages raise important questions concerning state tolerance of transnationalism, especially when "return migration" is not the goal. In these cases, a "stretching of the homeland" offers opportunities for diasporic peoples to own property, sponsor programs, attain education, and engage in the culture of their historic homeland without migrating or assuming the responsibilities of citizenship within that state. Those processes promise to bring people together but are complicated by governments with divergent positions on migrant rights. Regardless, diasporas have greater political weight than ever before in the contemporary international system.

The historical legacy of the British Empire in South Asia continues through transnational social fields occasioned by the steady flow of immigrants from India and Pakistan to the United Kingdom. On one hand, these migration streams provide a valuable source of labor possessing the language skills and cultural familiarity to contribute to British society, as well as adding another layer to an already cosmopolitan cultural landscape. On the other hand, however, postcolonial antagonisms and geopolitical tensions between India and Pakistan have been reproduced among their respective émigré communities. In this case, postcolonial migration helped spread existing geopolitical and border disputes between India and Pakistan and their associated ethnic tensions to British cities. This highlights the importance of understanding how new borders obtain their significance from the identities carried within individuals and groups. It also raises important questions concerning the value of border permeability. The crossing of borders has the potential to bring great benefits to states, provinces, and municipalities, but such crossings also have the potential to reproduce various forms of socioeconomic inequality, political rivalries, and even violence.

Insurgents and terrorists

War has traditionally been understood as a military conflict between two sovereign states, like the Iran-Iraq War or the Falklands War between the United Kingdom and Argentina during the 1980s. Yet most contemporary wars have pitted states against non-state combatants or alliances/coalitions of states acting against regimes deemed to have violated international law or norms. These types of efforts are generally sanctioned by the international community to some degree. NATO-led operations in Afghanistan and Libya are two recent cases, as is the UN-authorized coalition in the Gulf War in 1990–91. Other conflicts involve groups of states conducting military operations without broad international approval, such as the US-led invasion of Iraq in 2003 and repeated foreign incursions into the Democratic Republic of Congo.

Despite the dramatic nature of large-scale military operations, contemporary conflict is more likely to involve smaller-scale border crossings by non-state combatants. Quite often, these conflicts appear confined within a single state. Terms such as civil war, guerilla war, and domestic strife generally refer to violence between competing armed factions within the borders of a state, usually a government and an insurgent group, such as the Ethiopian government's conflict with the Tigray region, Yemen's battles with the Houthi movement, and Middle Eastern governments combating assorted branches of the Islamic State (ISIS). The domestic nature of these conflicts is misleading, though, since most insurgent groups operate beyond the borders of their home state. Assassinations, bombings, supply bases, training camps, fundraising networks, and other operations usually extend into neighboring or distant states. The fighting between Ukraine and Russia between 2014 and the full-scale invasion in 2022 illustrates the increasingly hybrid nature of armed conflict. Russia's annexation of Crimea in 2014 resembled a conventional military clash between two sovereign states, but the Russian infiltration was spearheaded by Russian soldiers wearing masks and uniforms without insignias who were already stationed in the region. Simultaneously, nominally independent secessionist movements among ethnic Russians in eastern Ukraine clearly received support and coordinated closely with the Russian government until Russia launched a full-scale invasion. Russia launched one main thrust of that invasion through Belarus using troops purportedly in that country for joint training exercises, but the exercises are widely regarded as part of Russian President Vladimir Putin's long-term plan to capture the Ukrainian capital of Kyiv and topple the government.

The Kurdish Workers Party (PKK) illustrates the cross-border nature of most insurgent movements. For decades, the PKK has fought to create a Kurdish state out of the territory of contemporary Turkey, including attacks in Turkey and against Turkish interests abroad. The PKK draws support from the Kurdish diaspora community and has training camps in northern Iraq and Iran. The US-imposed no-fly zone over northern Iraq inadvertently provided sanctuary for PKK bases and developed into a major source of tension. In response, Turkey refused to allow the United States to use Turkish territory during the 2003 invasion of Iraq. Turkey has subsequently violated Iraqi airspace on several occasions to target PKK forces and even invaded Iraq briefly in February 2008. This case further demonstrates the importance of understanding contemporary "civil" conflicts within a broader international context. Modern insurgent, revolutionary, and secessionist movements are rarely confined within a single state. Most involve some type of cross-border activity, thereby blurring the line between civil and international conflict.

Similarly, global terrorist networks undermine basic premises of the modern state system by relying on cross-border funding and communication to circumvent state security agencies. Terrorists are generally prosecuted in state legal systems like common criminals, though the scope of criminality extends far beyond the borders of any single state. The killing of innocent people for political purposes, as in the case of terrorism, commonly leads to murder charges in the state where the attacks occurred. Yet the preparations for that crime likely extended abroad. The individual may be funded and instructed to carry out the act by an organization operating in a different country or perhaps in the murky spaces of the internet. The prevention of future terrorist attacks normally involves actions and coordination across several different states. Such transnationality ironically compels states to further undermine international norms by launching preemptory or retaliatory attacks that transgress sovereign state borders, such as the 2021 US drone strike targeting ISIS members in Kabul, Afghanistan, that instead killed ten civilians, including seven children.

The Al Qaeda organization is the most notorious of recent terrorist organizations. Al Qaeda has a rather small membership, but its "brand" was embraced by various groups around the world to gain credibility and visibility for their causes. Instead of defending some homeland and patrolling borders, these terrorist groups function as a collection of loosely connected cells operating outside of the territorial norms of the nation-state system. The de-territorialized threats of these "non-state actors" compelled US President George W. Bush to declare war on the very tactic of terror. Unique in world history, the declaration raised the prospect of a "forever war" since a tactic cannot surrender or be defeated with finality.

The conflict along the Afghanistan-Pakistan borderlands demonstrates the complexity of the war on terror through its combative and cooperative intermingling of state militaries, insurgent militias, international coalitions, and terrorist movements. An American-led NATO force battled Al Qaeda and its Taliban allies in Afghanistan from 2001 until 2021. Following the withdrawal of US forces, many Taliban fighters taking refuge in northwestern Pakistan returned to Afghanistan. Although part of Pakistan, those "tribal territories" enjoyed considerable autonomy. Because they shared ethnic and cultural ties, many of Pakistan's tribal leaders were sympathetic toward the Taliban. These areas soon evolved into a sanctuary for Taliban forces to launch guerilla attacks in Afghanistan or for Al Qaeda and its successors to coordinate terrorist operations abroad. American forces were officially forbidden from crossing the border but nonetheless launched several missile attacks from drones patrolling Pakistani airspace.

These cross-border operations climaxed with the helicopter incursion of American commandos deep into Pakistan that resulted in the death of Al Qaeda leader Osama bin Laden in 2011. Human rights activists, international lawyers, and national security experts debated the morality and legality of the bin Laden operation and the continuation of drone strikes in Afghanistan and Pakistan. The murky criteria for determining whether groups or individuals should be classified as insurgents versus criminals highlight the complex nexus of cross-border terrorism, insurgencies, and crime. Regardless, the United States appears to have expanded its drone program in response to civil strife in Libya, Somalia, and Yemen, as well as ISIS fighters in Iraq, Syria, and other countries.

Criminals and police

Most states possess legal systems that base jurisdictional authority on territory. The borders between various national, provincial, and local jurisdictions help ensure the consistent and efficient enforcement of laws, punishment of criminals, and compensation for victims. Nevertheless, the idea of territorial sovereignty and the borders that demarcate jurisdictions often hamper efforts to stop criminal activity both internationally and domestically. Despite their intent to create and enforce some basic norms, legal systems and law enforcement remain fragmented, diverse, and in some areas largely absent.

Legal systems usually encompass a foundation of legal thought to define criminal behavior, a police system to enforce the laws developed from that foundation, a court system to apply the law, and a corrections system to either punish or reform criminals. In some countries, legal systems are highly centralized, so jurisdictions have little practical effect. Other countries have more decentralized law enforcement, so jurisdictions can make a significant difference. In the United States, each of the fifty states, as well as various local governments, has the right to legislate distinct codes of law and maintain separate police forces in accordance with the Constitution and Supreme Court decisions. This results in considerable differences in law across a wide range of topics, including abortion, firearms, marijuana, and same-sex marriage.

Despite efforts to institutionalize universal principles of human rights and international behavior, the modern state system lacks a police force or supreme authority to judge violations, enforce rules, and impose punishments. Instead, it relies on state law enforcement agencies to cooperate and share information in their efforts to combat cross-border criminal activity. INTERPOL and EUROPOL serve to coordinate extradition procedures and investigative cooperation among member states, but otherwise have rather limited authority.

Although not without controversy, the early twenty-first century has seen individual states formulate a variety of new legal procedures to address the realities of cross-border crime. Extradition and investigation were once the prime drivers of regular cooperation in the realm of international criminal justice, but new policies, agencies, courts, and detention facilities have proliferated in response to new threats. For instance, the United States has moved toward unprecedented intelligence sharing between domestic law enforcement agencies like the FBI, foreign intelligence agencies like the CIA, and foreign governments, such as the Five Eyes program that enables intelligence sharing between Australia, Canada, New Zealand, the United Kingdom, and the United States. These exchanges of information blur the legal distinctions between domestic law enforcement and international security activities and have drawn stark criticism from privacy activists, as well as "whistleblowers" who risk being convicted for treason like former National Security Agency contractor Edward Snowden. The Department of Homeland Security is itself a border-related agency whose purpose is to ensure the barrier/filter functions of US boundaries and facilitate the cooperation and implementation of best practices across domestic agencies.

Despite these efforts, cross-border crime remains broad in scope and difficult to interdict. Various forms of cross-border trafficking and smuggling are pervasive and utilize every conceivable mode of transportation, from cars, boats, and airplanes to tunnels, footpaths, and river fords. The definition of "contraband" varies by jurisdiction but encompasses a broad range of items. The diversity of things and means to smuggle create an exceedingly difficult task for law enforcement. Ships pose a particular problem relating to questionable arms transfers. Although ships are relatively easy to track, their cargo is much harder to police. Acting on intelligence from the United States, the Spanish navy intercepted an unflagged North Korean freighter shipping Scud missiles to Yemen in 2002. At the time, North Korea was designated as a state sponsor of terrorism and, although a US ally, Yemen was rife with political and ethnic strife. The United States disapproved of the weapons transfer but reluctantly allowed the delivery since, according to international law, Yemen was a sovereign state legally entitled to purchase conventional weapons, and the freighter was considered the sovereign "territory" of North Korea.

Unlike arms trafficking, drug smuggling can be accomplished through more modest means. Smaller boats, vehicles, and planes are often sufficient for the task. Some drug traffickers have even used "mules" who transport drugs on their person, sometimes even ingesting the drugs, and have proven especially challenging for law enforcement. Estimates vary, but studies agree that most illicit drugs trafficked across international borders are not interdicted and thereby make the leaders of drug cartels incredibly wealthy and powerful. In Mexico, rivalries and violent clashes between competing cartels and with government forces suggest that the Mexican government has lost effective control of portions its territory with violence resulting in over 120,000 deaths with another thirty thousand missing between 2007 and 2020. This tragic situation highlights the connection between drug and weapons trafficking as Mexican cartels utilize drug profits to purchase weapons and bribe government officials and border authorities, thereby expanding both the volume of contraband and accumulation of wealth. Globally, the nexus of drug trafficking and weapons smuggling helps fund insurgent groups and muddles the distinction between criminal cartels, political insurgencies, and terrorist networks.

The related enterprises of human smuggling and trafficking facilitate illegal flows of people across borders. Human smuggling involves assisting a person to gain unauthorized entry into a foreign country, usually in exchange for payment. The assistance could entail guiding migrants across the border clandestinely or providing fraudulent documents to pass through an official entry point. Although it takes different forms ranging from temporary bonded servitude to forced labor, human trafficking refers to the actual trade in human beings as property and is often termed "modern-day slavery." The people bought and sold are generally young adults or teenagers, or even younger children, who are exploited for work as domestic servants, sweatshop laborers, soldiers, or sex workers. The process of trafficking does not require crossing an international border, but in practice most human trafficking operations have some cross-border dimension.

Other illicit activities, such as money laundering in offshore banks, are complex financial crimes that require the crossing of borders to take advantage of more lax regulations in different territorial jurisdictions. Such efforts follow similar practices of extra-territoriality that enabled the circumvention of US laws for enhanced interrogation of terrorists, but in the financial realm, certain banks in Switzerland and the Cayman Islands, and even the US state of South Dakota, afford opportunities for establishing accounts or perpetual trusts that enable tax evasion or money laundering. Real estate in global hubs such as London and the high-end art market are other mechanisms of financial "off-shoring" that allow for the safe "parking" of wealth with little scrutiny.

The infringement of intellectual property and patent rights is equally complex and seems to occur with near impunity in many places. Those "pirating" movies, music, and other forms of intellectual property should technically be arrested and prosecuted by state authorities. However, international borders reduce this possibility. Chinese open-air markets are typically replete with counterfeit goods bearing the brands of prominent companies. Marketing such "knock-off" goods without the permission of the patent holder is a crime, but the prospects of enforcement are slim due to state sovereignty.

Copycat pharmaceuticals distributed in developing countries are another example of intellectual property rights infringement that carries complicated ethical considerations. Although resulting in less expensive medicines that are more accessible for poorer people, patent-holding companies will have diminished global market share and revenues to fund future research and development. Consequently, the development of new treatments and medicines for a variety of diseases and medical conditions may be slowed or stopped entirely. Even if a reliable enforcement mechanism existed, the prosecution of companies replicating patented drugs without approval raises the ethical quandary of restricting access to medicines in impoverished regions.

Other cross-border crimes may not have those ethical considerations but are equally murky. Computer hackers prowl the virtually unbounded realm of cyberspace to wreak havoc on public and private computer systems. Hacking was initially associated with young pranksters, but more recently powerful government agencies and sophisticated nonstate actors have engaged in hacking to disrupt various economic, social, and political processes. In September 2021, the European Union formally blamed Russia for its involvement in the "Ghostwriter" cybercampaign, which targeted the electoral and political systems of several member states. Since 2017, Russian operators hacked the social media accounts of government officials and news websites with the goal of creating distrust in US and NATO forces. Such state-to-state attacks are hard to prove but seem to have become just another modality of international politics.

Similarly, nonstate actors seeking to profit by penetrating the cyber borders of both companies and private citizens are so common that most businesses equip their workers with VPNs and dual security verifications. Any breaches raise the prospect of blackmail or identity theft, as in the May 2021 ransomware attack on the largest fuel pipeline in the United States or the August 2021 leak of personal details from more than fifty million mobile phone customers. In contrast to these nefarious acts, an August 2021 hack of a high-profile prison in Iran uncovered documents, videos, and images illustrating the inhuman treatment of inmates. While no less a violation of bordered cyberspace, such acts in support of human rights, environmental, or humanitarian causes have been deemed the work of "hacktivists." Following Russia's invasion of Ukraine, a Twitter post from an account named "Anonymous"— with 7.4 million followers and nearly 190,000 retweets— summoned hackers around the world to target Russia. Programmers within and beyond China are also developing apps to circumvent the censoring capacities of the so-called Great Firewall of China, demonstrating that cyber borders invariably trigger efforts to work around those restrictions.

The internet has proved to be such a powerful social force and so integral to socioeconomic and political life that many governments have developed "kill switches" that allow them to shut down cyberspace within all or parts of their territory. China has cut off internet connections in the Xinjiang region to conceal systemic government oppression of minority Uyghur populations, while India routinely shuts down internet and cell-phone service in Kashmir and adjacent regions to quell antigovernment protests. The Iranian government also activated its kill switch to prevent organized opposition and riots during protests in 2019–20. Even politicians in the United States proposed legislation to provide the president with an internet "kill switch" should a cataclysmic cyberattack threaten energy, transportation, or other vital systems, although the proposal received intense criticism and was ultimately unsuccessful.

Tourists and travelers

News reports tend to focus on the negative aspects of border crossings—real, potential, or imagined—but it is important to emphasize that border crossings are generally positive in nature. Tourism—or travel for the purpose of leisure—is one of the most prominent and positive forms of border crossing. In addition to fun and relaxation, tourism offers opportunities to gain a deeper appreciation for new places, peoples, and experiences. Tourism also provides a major pillar for global economic development, estimated to account for approximately 10 percent of all jobs and 10 percent of global gross domestic product before the COVID-19 pandemic. Obviously, tourism can greatly benefit the traveler and the travel destination, as well as assorted businesses that provide travel services, but the potential negative side effects and corresponding ethical questions should not be overlooked.

Tourism is commonly defined in terms of distance traveled, places and people visited, and time spent away from home. It invariably involves crossing borders. These borders could be municipal, county, provincial, or international, but the crossing of any border feeds into the experiential side of tourism and the idea of leaving your daily routine and territory. For many, the excitement of the journey is enhanced by crossing highly visible political or cultural borders. Tourism studies have shown that border crossing adds to a trip's perceived distance and the more draconian the process of traversing the border the greater the perceived distance. Travel within democratic, developed states is generally less arduous, but the required travel documentation within less developed or nondemocratic states may be as extensive as international trips.

Whether motivated by leisure or business, travelers crossing international borders are subject to scrutiny by both the country of origin and destination symbolized by the documentation required to prove individual identity, place of citizenship, and by extension one's rightful place in the world. Historically, travelers carried simple letters of introduction, but the emergence of nation-states led to new rules governing international travel. Gradually, the passport gained acceptance as the most trustworthy proof of identity. Modern passports and related identity documents are increasingly loaded with personal travel and citizenship data, anti-counterfeiting technologies, biometric tracking capabilities, and most recently requirements for health information such as vaccination records. Requirements for identity documents to legally cross borders have steadily become more extensive.

Once they satisfy these bureaucratic requirements, tourists often seek out markers of territorial difference by posing for photos next to signs, fences, walls, or other types of boundary markers. Some borders have actually become tourist destinations in and of themselves. Niagara Falls (United States/Canada), Victoria Falls (Zambia/Zimbabwe), and Iguazu Falls (Brazil/Argentina) combine stunning natural attractions and international borders. Relict borders such as the Berlin Wall, Hadrian's Wall, and the Great Wall of China attract thousands of people annually. Even the odd meeting points of multiple borders have become popular tourist destinations, such as the Four Corners Monument where Arizona, Colorado, New Mexico, and Utah converge, the Tri-point

Monument where Finland, Norway, and Sweden meet, and the Triple Frontier of Argentina, Brazil, and Paraguay.

Regardless of the journey's purpose, all travelers must navigate different sovereign legal regimes as they cross borders, which may result in gaining or losing certain freedoms or restrictions. Some of these might involve drastic differences in basic personal freedoms pertaining to mobility, speech, religion, or sexuality, while others involve the accessibility of certain goods and services, such as alcohol, pharmaceuticals, gambling, or prostitution. Tourist destinations often market these different legal regimes to attract specific tourist populations, while tourists commonly exploit different currency zones to enhance their purchasing power, especially in borderland regions. Medical tourism has emerged as a prominent means of gaining access to treatments or surgeries either unavailable or unaffordable in the home state. In countries with *jus soli* citizenship laws that confer citizenship automatically upon those born within the state's territory, birth tourism affords an "anchor baby" citizenship that can thereby help the parents attain legal residency.

Catering to tourist desires and expectations may also risk commodifying and trivializing local cultures and thereby lead to questions of cultural authenticity and appropriation. Tourist destinations that become conspicuously dependent on cross-border flows could also be seen as "foreign" to that local and national culture. Some communities are increasingly concerned with "over-tourism" and the erosion of traditional place identities resulting from the influx of short-term tourist visits and longer-term vacation rentals. In response, some governments have restricted the operations of Airbnb and similar tourist businesses to established tourist zones in order to maintain quality of life and daily routines for locals in adjacent neighborhoods. Although it was broadly suspended during the COVID-19 pandemic, tourism rapidly rebounded as a growing economic activity and ubiquitous experience of border crossing.

Chapter 6
Cross-border institutions and systems

Cross-border institutions and systems are today far more pervasive and central to the lives of more people than in any previous era. The existence of these cross-border networks undermines the common assumption that the borders of state, society, and economy are, or at least should be, congruent. Environmental concerns, public health issues, and flows of information, for example, are only partially subject to the power of borders, yet borders clearly have an impact on the ecological, medical, and intellectual realties of human existence. Being ironically resistant to and profoundly influenced by political borders, cross-border institutions and systems reveal the inherent complexity of borders and border research.

Ideas and information

Thanks to new communication and transportation technologies, ideas and information cross borders with unprecedented ease, volume, and speed. The emergence of "cyberspace" and the global reach of media broadcasts facilitate massive flows of information and ideas and thereby challenge state sovereignty and territorial borders. The technologies may be new, but ideas and information have a long history of circumventing borders, often to the chagrin of governments. The British Broadcasting Corporation's World Service and the US government's Radio Free Europe and Voice of

America transmitted democratic ideals to audiences in communist countries during the Cold War. Even acknowledging these earlier precedents, it is important to state that contemporary cross-border exchanges of ideas and information are unprecedented, with profound and contradictory implications for both political and socioeconomic borders.

Modern telecommunication satellites may lack weapons but nonetheless represent a type of strategic asset through their ability to transmit information and influence public opinion. Western companies, such as the BBC, News Corporation, CNN, Deutsche Welle, and France24, dominate the satellite television broadcasting market, while Netflix, Amazon Prime, Apple Music, HBO, Hulu, Spotify, Sirius, and Disney+ are prominent in global streaming and therefore in key positions to shape cultural norms and political agendas. It is obvious that broadcasting "the news" from a particular ideological perspective holds the power to shape hearts and minds, so the dominance of western corporations over global media—film, television, and music—carries significant international implications. Regional outlets, such as Al Jazeera and Al Arabiya, have sought to counter western dominance among Arab-speaking audiences in the Middle East and the ethnic Arab diaspora, while the World Is One News similarly provides news coverage from an Indian perspective for domestic and international audiences. Yet, even these flows are subject to bordering as numerous states censor or block satellite communications.

New cyberspaces of human interaction are also forming as the internet enables novel exchanges of ideas and information. The seemingly boundless and borderless nature of cyberspace offers a vast array of connections for those with internet access. Internet communities provide venues for identity and belonging to transcend the territorial limits of state borders and sovereignty. The premise of "shared interests" has given rise to a wide range of new international communities. Some of these online communities are global humanitarian or environmental movements with specific ideological objectives, such as the International Committee of the Red Cross or Greenpeace. Other online communities are simply intended for socializing, entertainment, and leisure, such as dating services, gaming groups, or social media platforms. In most cases, these new linkages encourage constructive and healthy cross-border dialogues. Unfortunately, like all identities, even these internet communities can develop privileged and hierarchical structures with the potential for exclusion, bigotry, and even human rights abuses. The generally open nature of cyberspace facilitates the harmful cross-border actions of computer hackers, terrorists, criminals, and governments to suppress dissent and meddle in foreign elections, just as it serves the humanitarian aims of Amnesty International and Reporters Without Borders.

Some regard the internet as the primary agent in flattening the world and erasing state borders. Yet despite its apparent openness, there is a clear geography of internet censorship. Many of the nearly 4.6 billion internet users face government restrictions and surveillance as states invest considerable resources in controlling and monitoring online content and activity. Governments block websites, monitor chat rooms, and harass users of social media, among other limitations on cyber freedom. Democratic states debate the legality of implementing internet restrictions, raising difficult ethical issues concerning the proper balance between freedom of speech and preventing the distribution of reprehensible content, such as child pornography. Certain social media companies are under intense scrutiny as platforms for "influencers" with the capacity to shape elections, vaccination rates, and individual reputations through "fake news" and cyber bullying. Given the massive and decentralized nature of internet traffic, the effectiveness of such programs is unclear, but governments have taken a greater interest in regulating online flows of information and ideas, in some cases at the behest of some of their constituents.

The so-called Great Firewall of China exemplifies government cyber censorship. Beginning in 2003, China's communist government launched Operation Golden Shield to block access to certain internet sites, programs, and content. The project also censors any websites considered politically subversive by communist authorities, including websites advocating democratic reforms in China, independence for Tibet or Xinjiang, or human rights. China also blocks access to Facebook, X, formerly known as Twitter, YouTube, and several other social media sites, fearing they could be used to document human rights abuses or organize antigovernment protests. Under the leadership of President Xi Jinping, China has institutionalized extensive competitive advantages for domestic internet firms through regulation, censorship, and outright bans of foreign media and technology platforms. As a result, Chinese firms and websites like Baidu, WeChat, and Weibo dominate domestic markets.

China's censorship has been generally successful, but that is not the case for all governments. In January 2011, the reign of Tunisian dictator Zine el-Abidine Ben Ali abruptly ended after twenty-three years. The uprising that pushed this dictator from power was not, however, driven by an ideologically motivated insurgent group but rather by popular discontent and anger toward the government spread through the social networking website Facebook. After a police officer confiscated his produce cart, the young Tunisian Mohamed Bouazizi protested by setting himself ablaze in front of the regional governor's office. Tunisian citizens uploaded photos and videos of subsequent public dissent from cell phones to Facebook and succeeded in stirring the population to broader, sustained protests. In response, government authorities implemented a program to steal online passwords and login information, thereby enabling them to block further use of the social networking site, as well as potentially identifying those involved in the protests. Undeterred, protestors created new accounts to advance their cause. Eventually, the government succumbed to public pressure, and Ben Ali fled the country.

Military strategists, government leaders, and political dissidents have long recognized the importance of information and propaganda, but the overthrow of Ben Ali may be the first successful cyber revolution. More recently, social media proved integral in organizing the Black Lives Matter (BLM) protests of 2020 fueled by the murder of African American George Floyd by police officers in Minnesota that was recorded and disseminated through Twitter and other social media platforms. In combination with other instances of police misconduct against African Americans, a nationwide movement took form demanding accountability and overall police reform. The prominence of BLM protests in some areas and relative absence in other places highlighted geographical differences in social practices and cultural spaces.

Social practices and cultural spaces

Facebook, TikTok, X, formerly known as Twitter, and Snapchat make obvious the border-crossing capacity of cyberspace, but state borders continue to shape and reflect a variety of social ideas and norms. Any consideration of borders must acknowledge the lines that divide the everyday social practices of males and females. The most obvious of these are gender-specific spaces designed to separate the sexes, often reflecting different gender roles derived from religious customs. Assertions of gender fluidity have made even formerly benign bordered spaces, such as public restrooms, controversial. In certain cultures, segregation by gender attracts little attention, as in the case of Islam having separate spaces for women and men to worship. In other settings, gendered spaces have generated greater controversy. The Islamic custom of women wearing veils or other coverings, for instance, has sparked significant debate in many western countries. For some, the veil simply reflects the ideals of modesty, fidelity, and marriage. Others interpret the veil as a type of social barrier limiting women's rights and fostering discrimination.

Other recent trends involve attempts to export particular understandings of feminism, gender, and sexuality in western societies to other countries. Proponents tend to portray these efforts as promoting universal notions of human rights across international borders, but opponents counter that feminism, gender, and sexuality have different connotations in non-western cultures or even interpret this as simply the most recent iteration of western cultural imperialism. The Taliban's efforts to cordon off Afghanistan during the 1990s and reimposition of control following the US withdrawal in 2021 are unambiguous cases of borders as tools to prohibit foreign cultural influences.

Within Taliban-dominated Afghanistan, a wide variety of previously lawful activities and products are once again forbidden under a skewed interpretation of Sharia law combined with local tribal traditions. The prohibitions included products containing human hair, satellite dishes, certain musical instruments and audio equipment, pool tables, computers, VCRs, televisions, lobsters, dancing, kites, nail polish, firecrackers, statues, and sewing catalogs. Women will once again have very limited opportunities for employment, education, travel, healthcare, and sports under Taliban rule. For their part, men are expected to grow beards of a certain length, keep their head hair short, and don head coverings. As Afghanistan reverts to Islamic authoritarianism, the borders of the state become conspicuous in facilitating these extreme cultural restrictions. Like Stalin's efforts to impose an "Iron Curtain" across Eastern Europe, the Taliban seek to construct a "mud-brick wall" around Afghanistan within which their version of religious piety and purity can be realized. The result is a realm of oppression for the majority of the population. It also remains to be seen whether Afghanistan resumes its role as a haven for extremist groups, such as ISIS, as it once served for Al Qaeda.

Even in democratic western societies, borders remain common tools for those seeking to promote various forms of differentiation or even discrimination. Sub-national borders often create and perpetuate social categories and hierarchies rooted in complex histories. Voting districts, census tracts, and school zones serve as conduits for socioeconomic status and political power or the lack thereof. States subdivide sovereignty and jurisdiction in various ways, but every country allocates some degree of power to subordinate governments. The borders of local, municipal, provincial, and federal entities mark off spaces of responsibility and institutionalize levels of jurisdiction. These hierarchies of places structure modern societies and, some would argue, constitute the basic building blocks of contemporary democratic practices. Indeed, it is unclear whether democracy on any meaningful scale can form and function without clearly defined territories and jurisdictions.

Most democratic systems feature domestic borders designed to ensure territorial, as well as demographic, representation, but the processes of forming such legislative districts are highly vulnerable to deliberate manipulation for the purpose of influencing elections—i.e., gerrymandering. This said, proof of residency or citizenship are common requirements to ensure equal voting rights within jurisdictions ranging from municipalities to states. Reflecting British Prime Minister Winston Churchill's statement that "democracy is the worst form of government except for all the others," the risk of corruption or historically institutionalized inequalities that limit legitimate democratic representation of ethnic, racial, or economic groups remains prevalent. Nevertheless, the democratic process requires setting some scope for popular participation and representation. In other words, state, provincial, district, and municipal borders help set practical and manageable parameters for polling the electoral body, as well as defining the jurisdictional realm of the elected officials and governments. Modes of defining citizenship vary, but the vast majority remain circumscribed by borders, whether in the mode *jus soli* (citizenship connoted through birth within a territory), *jus sanguinis* (citizenship determined by parental citizenship), or *jus nexi* (citizenship earned in place).

Clearly, democracy as a practice of governance has come to be inexorably bound with territorial sovereignty. The transition from religious-monarchial sovereignty to popular-territorial sovereignty was facilitated by the unity of the people (real or imagined) and their sanction (real or imagined) of a representative government. Following the American and French Revolutions, territorial sovereignty and popular democracy grew in tandem. Nationalism reinforced the linkages between those ideals through constructed histories of social unity, common struggles, and shared sacrifices, which symbolically rooted the ethos of the people within a particular territory. The rapid proliferation of this framework over the past two centuries solidified the relationship between democracy and territory, as well as demarcating the borders between contemporary nation-states. These linkages have become so embedded that criticizing the notion of territorial sovereignty is commonly interpreted as challenging the rule of "the people." Standard political maps naturalize this relationship by portraying the world as a collection of separate territorial units. Yet, this obscures the complexity of cross-border relationships and the daily practices of integration that pervade the contemporary international scene.

Supranationalism and regionalism

The development of supranational organizations also challenges the territorial assumptions of the modern state system. Supranationalism refers to the process of states transferring some portion of their sovereignty to larger quasi-federal organizations.

Some of these supranational organizations play highly visible roles in international diplomacy and trade, such as the United Nations and World Trade Organization. For a time, it appeared that supranational organizations might take the place of sovereign states and fulfill the same basic role but at larger scales. For some, the emergence of supranational organizations portended a cosmopolitan democracy where global citizenship and universal human rights would replace the current world of territorial belonging and state sovereignty. Most proponents of "no borders" or open borders envision such a scaling up to global democracy, although it is unclear how that would happen exactly.

The evolution of Europe from the tinderbox of two world wars to the standard-bearer for supranational integration to the recent discord among members epitomized by Brexit has special significance for the dynamic relationship between supranationalism and borders. Rooted in the desire to preclude a future war by promoting economic integration and cooperation and addressing cross-border problems like pollution and crime, the European Union facilitated successive treaties between member states establishing freedom of movement for goods, services, capital, and people among an expanding number of member states. Effectively pooling sovereignty at a continental scale, the European Union seemed to herald a realignment of the state system that promised to make territorial sovereignty increasingly irrelevant. The adoption of the Euro common currency, visa-free travel within the Schengen Regime area, standardized passports for external travel, and the ability to vacation, study, and work across the EU's territory seemed to suggest bordered nation-states might come to function as substate units with a larger federal structure.

Ironically, even as it appeared to be moving inexorably toward open borders among its members, the European Union simultaneously launched several initiatives aimed at hardening its external borders. Today, EU member states bordering non-EU states are required to demonstrate efficient policing of their external borders and are eligible to receive EU funds to strengthen their ability to block crime and illegal migration. Poland, Latvia, Lithuania, and other EU states blocking the entrance of migrants primarily from Iraq attempting entry through Belarus in late 2021 served as a vivid illustration of efforts to strengthen the EU's external borders. Russia's invasion of Ukraine and the complicity of Belarus in facilitating the onslaught will likely compel additional security measures along the EU's eastern borders. The European Union has even pushed border enforcement initiatives far beyond its actual borders to stem migration at source countries across Africa and the Middle East, raising the specter of "Fortress Europe."

The Great Recession that swept the globe in the late 2000s left several EU members unable to repay their debts without various types of loans, credits, guarantees, and "bailouts" from other members and the International Monetary Fund. This debt crisis triggered great acrimony and discord in creditor countries where people resented having to subsidize another country's debt and in debtor countries where people felt belittled and blamed. These bitter feelings rapidly resurfaced as Europe faced a surge in migration during the 2010s largely propelled by refugees fleeing military conflicts in Afghanistan, Iraq, Libya, and Syria, as well as widespread poverty across much of Africa and Eurasia. Populist movements proliferated across Europe demanding that governments reimpose border controls and reassert nation-state sovereignty in general. Governments soon recognized the rapidly eroding support for European integration or faced the prospect of being swept from power.

The successive backlashes against the debt crisis and immigration undermined notions of European solidarity. Although often announced as temporary measures, the return of rigorous border controls exemplified by a spate of construction of walls, fences, and detention facilities may mark the beginning of the region's re-bordering. The decision of the United Kingdom to leave the European Union—the so-called Brexit—provided an even more dramatic turn of events with permanent implications for cross-border integration in Europe.

The changing status of the border between Northern Ireland, the United Kingdom, and the Republic of Ireland, an EU member, is a poignant case of this re-bordering. Since the 1998 Good Friday Agreement, the border between the Republic of Ireland and Northern Ireland had been largely open to commerce and travel, but Brexit promised the reintroduction of border controls. In 2021, a protocol was enacted that allowed free trade to continue across the border, effectively treating Northern Ireland as if it was still in the European Union. Ironically however, this agreement impedes trade between Northern Ireland and the rest of the United Kingdom because EU regulations require the inspection of certain goods arriving from non-EU countries. Products such as milk, eggs, and frozen meat coming from England, Scotland, and Wales to the Republic of Ireland through Northern Ireland must be inspected in Northern Ireland's ports since entry into Northern Ireland affords free passage to the Republic of Ireland and the rest of the European Union—effectively making the Irish Sea a new border between Northern Ireland and the United Kingdom. Because either side may suspend the protocol should circumstances exist that "cause social, economic, or environmental difficulties," this situation could escalate should the European Union continue to reject the United Kingdom's petition to renegotiate the border protocol.

Fiscal crises, immigration, and crime are the foremost catalysts of new bordering trends in Europe, but the European Union has also embraced the principle of subsidiarity, which holds that governmental responsibilities should be handled at the lowest possible level of government. This principle has helped fuel a process of political decentralization as states such as Belgium, France, Italy, and Spain have gradually moved toward more

9. Protestors in Northern Ireland call for "secure" borders to block Muslim refugees.

federal systems of governance since the 1970s. Benefitting from greater autonomy, a new regionalism is taking shape as some provincial governments find cross-border exchanges, associations, and partnerships with neighboring regions more advantageous than with other areas of their nominal state.

The European Union's success in eliminating so many barriers between member states and promoting subsidiarity may ironically work to make independence movements more viable. It now appears increasingly feasible for smaller regions to gain independence while still retaining the economic and security advantages of EU membership. The Padania independence movement in northern Italy, which claims a cultural heritage distinct from southern Italian culture, is a prime example. In other places, such as Catalonia, Corsica, Flanders, and Scotland, the reemergence of older regional frameworks combines with strong ethnic identities to challenge the territorial integrity of their respective states. Brexit and its wide-ranging and still uncertain consequences similarly call into question the territorial integrity of the United Kingdom.

The European Union is one of the most visible and developed supranational organizations, but many others exist and provide important forums for cross-border cooperation. Many of these institutions focus on economic issues and trade, such as the Asia-Pacific Economic Cooperation forum or Eurasian Economic Community. Others focus on security concerns and military cooperation, such as the Collective Security Treaty Organization or the North Atlantic Treaty Organization. Still others, like the Africa Union or Union of South American Nations, strive for deeper and broader cooperation modeled on the European Union. These organizations have generally been most effective in areas concerning economic cooperation, especially liberalization of trade, but it remains uncertain whether lasting cooperation or greater integration will emerge on other issues.

Despite a general media focus on economics, much of the popular support for supranationalism around the world derives from social welfare issues, human rights, and environmental concerns. The early stages of European integration drew impetus from efforts to combat water and air pollution, and those are areas where the European Union has earned some of its highest approval ratings. Domestic and nonstate organizations increasingly stand at the forefront of environmental causes and proclaim the simple truth that political borders rarely correspond to the geographical realities of pollution, natural hazards, and ecosystems.

Environmental concerns and borders

Borders are human constructs, but the natural world is also divided into distinct habitats, biomes, climatic zones, and ecosystems. Yet the frontiers between those natural phenomena tend to be more like fluid and ever-changing transitional zones than clear border lines. Borders would seem to have little bearing on nature, and when borders do play a role, they are more likely to seem as hinderances to addressing environmental challenges. The massive locust swarms that threatened parts of Ethiopia, Kenya, and Somalia in 2019–20 aptly demonstrated that no single state could combat the plague, although all bore the cost as the insects wreaked havoc on crops. The intrinsic cross-border character of natural phenomena complicates assumptions of territorial sovereignty that constitute the foundation of the contemporary state system.

Perhaps in recognition of nature's indifference to political borders, governments are increasingly using borders to promote environmental conservation. Nature preserves and national parks represent efforts to safeguard natural settings by creating bordered spaces that exclude certain human activities. Established in 1872, Yellowstone National Park in the western United States is generally recognized as the first modern national park. Many other states have since established their own park programs, often blending environmental conservation with more parochial goals of nationalizing and politicizing state territory. But even with the best of intentions, it is practically impossible to completely cordon off nature preserves and parks. At the very least, seasonal migrations of animals and insects and seeds carried by wind and water will freely cross borders. As a result, states increasingly recognize the advantages of cooperative approaches to environmental conservation. The formation of the Kgalagadi Transfrontier Park across the borders of South Africa and Botswana and La Amistad International Park between Costa Rica and Panama exemplifies this emerging trend.

Unfortunately, borders can also hinder conservation efforts since states with lax or nonexistent environmental regulations are still recognized as sovereign and therefore entitled to manage their territory free from outside interference. In response, some international organizations, like the World Wildlife Fund and other environmental nongovernmental organizations (NGOs), have tried to intervene in support of creating cross-border parks and preserves for endangered species. By seeking to impose their own preferences across state borders, these NGOs may undermine local traditions or government policies, but the notion of environmental interventionism appears to be gaining acceptance. There are currently movements to institutionalize global standards governing fish and wildlife harvests, bioreserves, water pollution, carbon emissions, hazardous waste disposal, and the testing of nuclear weapons. "Climate initiatives" conspicuously abound at the regional, state, and global scales.

These appear to be noble objectives, but each raises complex cross-border issues. Whether motivated for ecological or economic reasons, bordering as a mechanism for environmental protection spans emissions controls, fishing and hunting regulations, categories of endangered and threatened species, and timber harvest limits and replanting requirements. These policies have legitimate purposes but may be manipulated to serve specific interest groups. Massive Siberian wildfires in 2021, for instance, were rumored to be the result of arsonists seeking to alter the status of protected forests in response to a statute that allowed logging within burned acreage. Like any act of bordering, even those seeking to protect the environment are inherently political.

Taking a broader view of borders and environmental sustainability, nature has its own spatial logic. As such, increasing human mobility and our capacity to transport species across space constitutes a prevalent driver of environmental change and has risen to the forefront of domestic politics and international relations. The infamous introduction of the rabbit and the cane toad into Australia disrupted regional ecosystems, while the United States has seen similar problems stemming from invasive species, such as kudzu plants, Japanese knotweed, zebra mussels, Asian carp, and even "murder hornets." These invasive species have triggered extensive containment efforts and restrictions on the transportation of species. Unfortunately, these efforts have had limited success. On a more benign note, the breeding of pandas in zoos in both the United States and China from the 1970s served as launching points for higher-level diplomatic relations. These types of conservation efforts can be integral tools for promoting cross-border cooperation, especially concerning the management of transboundary natural resources.

Borders can also complicate the management of natural resources, especially transborder water resources. The Colorado River, for example, originates in the US state of Colorado, yet much of the water is diverted for various uses as it flows downstream before the river reaches "lower basin states" like Arizona, California, and Nevada or crosses the border into Mexico. The major river systems in Central Asia span several states, but these states also have competing priorities for utilizing the water. The upstream states of Kyrgyzstan and Tajikistan want to release reservoir water through hydroelectric dams during winter to generate heat and power. The downstream states of Uzbekistan and Turkmenistan want water released in the summer for irrigation. Similar tensions persist between Israel, Jordan, Syria, and the West Bank regarding the Jordan River. These cases raise questions as to ethics, natural resource usage rights, and territorial sovereignty.

Health and borders

By defining the limits of state sovereignty and domestic jurisdictions, borders delineate spaces marked by profound differences in standards of living and quality of life. Indeed, rights to employment, housing, travel, healthcare, education, self-expression, and even family size vary significantly from country to country and jurisdiction to jurisdiction. Some countries have constructed elaborate systems to protect their citizens by regulating things like guns, toys, appliances, food, medicines, tobacco, and alcohol. The US Customs and Border Control lists a variety of items that are illegal to bring into the country, famously including Cuban cigars, as well as other products from embargoed countries, various medications, and certain seeds and soils. The United States is rather strict on such issues, but other states do nothing or very little in this regard.

Health is among the most variable aspects of quality of life, and international borders play a substantive role in shaping those differences. Almost every imaginable facet of health exhibits significant variation from state to state, including fertility and mortality rates, life expectancy, and access to professional healthcare. Life expectancy, for instance, averages around eighty years in many developed countries but only around forty-five years in some of the poorest countries. Beyond differences in life expectancy, the actual causes of mortality also vary significantly from country to country. Mortality in developed countries is mostly attributable to a combination of lifestyle choices, like tobacco use or physical inactivity, and problems associated with old age, like heart disease or cancer. In contrast, major causes of mortality among populations in developing countries often include relatively basic medical problems, like respiratory infections or dysentery often worsened by malnutrition. Simply put, borders divide and create spaces and populations of vastly different health problems and outcomes.

The notion of environmental justice signifies uneven geographic exposure to adverse health effects from air and water pollution and a range of other environmental risks. Even on a local scale, exposure to environmental problems and their associated health risks is disproportionately concentrated in areas of low-income, minority, or indigenous populations. For example, polluting factories are more likely to be located near low-income neighborhoods. On a global scale, the harmful impacts of overfishing or logging are most severe in poorer, developing countries. This dynamic is even found in global recycling flows as products containing toxic materials, like electronics, chemicals, plastics, and even cargo ships, are more commonly exported to developing countries for processing.

Numerous other health issues cross state borders. As became abundantly clear in March 2020, disease is an obvious transborder issue. An epidemic signifies the extensive spread of a contagious disease across social and domestic boundaries, while a pandemic signifies that the disease has spread globally. Globalization and increased mobility of people provide new vectors for disease diffusion at much greater speeds. Cholera originated in South Asia, for example, but developed into a global health threat that has periodically wreaked havoc on populations for centuries, most recently in Yemen from 2016–21.

HIV/AIDS has also diffused globally without regard to international borders and social boundaries but is clearly more prevalent in impoverished regions. The disease continues to impact many parts of Africa, where programs designed to slow the contagion are hampered by conspiratorial rumors, long-standing social norms, and overall lower standards of healthcare. The states hardest hit by this disease have limited capacities to prevent its spread into or out of the country. The COVID-19 pandemic, generally believed to originate from Wuhan, China, has affected the entire world. The pandemic highlighted differences across the globe and within national territories regarding the capacity to provide treatment, political perspectives on disease mitigation, regional shortcomings in hospital infrastructure, and varied accuracy in identifying and calculating infection rates.

The preference of some governments to allocate scarce resources to the military rather than healthcare demonstrates how territorial sovereignty contributes to difference in health outcomes. A more complex example is the varied legality of certain medicines, abortion, or assisted suicide. Each of these policies is determined by state law. It is for this reason that borders remain an integral force for producing uneven healthcare practices and outcomes around the world. Numerous organizations, such as Doctors Without Borders or the World Health Organization, are border-crossing organizations with the express purpose of redressing the disparities in human health and quality of life that have been institutionalized by international borders and the ideal of territorial sovereignty.

Such disparities are not, however, unique to the international sphere. Internal jurisdictions and socioeconomic borders are equally influential in shaping differences among domestic populations. Residential segregation based on socioeconomic status is common in many countries. This creates spatial patterns where socioeconomic differences between adjacent neighborhoods or communities engender and perpetuate differences in overall health and quality of life. It may fairly be stated that political and socioeconomic borders, both international and domestic, create profound differences in overall well-being and opportunity, including access to healthcare, educational opportunities, exposure to pollution, and availability of clean drinking water and sewage systems.

Ethics and borders

Given their role in shaping stark disparities in quality of life, borders raise numerous ethical issues and are inseparable from moral assumptions and judgments. Societies since antiquity have possessed general codes of conduct, but the actual prescribed and proscribed practices have varied greatly. Traditionally, these moral concerns derived from specific religious belief systems. Animist, eastern, and western theologies developed their own specific customs related to territory and borders ranging from procedures for property ownership within religious communities to relationships with lands governed by other religious groups. Colonialism provided a mechanism that propelled European territorial practices and state conduct into global norms and conventions. Initial efforts during the nineteenth century, such as the First Geneva Convention, tended to focus on state responsibilities and acceptable conduct during wartime, such as the treatment of prisoners of war and civilians. These conventions were largely ineffectual as the development of "total war" during World War II led to massive civilian casualties, expulsions, and genocide.

These tragedies provided renewed impetus to develop and apply broad-reaching conventions of international law and human rights. In addition to affirming the centrality of state territorial sovereignty, the founding charter of the United Nations also listed "promoting and encouraging respect for human rights and for fundamental freedoms for all" as one of its main purposes. In 1948, the United Nations issued the Universal Declaration of Human Rights (UDHR) to clarify the meaning of "human rights" and "fundamental freedoms." The UDHR set out a concrete list of human rights, but it was not a binding treaty and lacked monitoring and enforcement mechanisms. Instead, the observance of human rights was left to states, which often prioritized geopolitical or cultural agendas that ran counter to the lofty ideals embodied by the UDHR.

Many contemporary scholars and activists have since emphasized how the continuing primacy placed on territorial sovereignty and border enforcement often facilitates assorted human rights violations. States accused of human rights abuses commonly assert the principles of state sovereignty and territorial integrity to block intervention. This impasse in contemporary international law has led some to argue that ethical and moral concerns, specifically protection of basic human rights, justify cross-border interventions. In this view, protecting basic human rights takes precedence over state sovereignty.

Events in Darfur, Sudan, illustrated the inherent contradictions between state sovereignty and human rights. Some governments and many human rights NGOs characterized the actions of the

Sudanese government since 2003 against the residents of Darfur as a campaign of genocide. Therefore, foreign intervention was justified to protect civilians and uphold human rights. The Sudanese government and its allies responded that these actions merely represent a sovereign state exercising the right to administer and police its territory free from outside interference. In this case and many others, judgments of what qualifies as an "international" crime against humanity versus the legitimate actions of a sovereign state are often secondary to other political and economic considerations.

The inconsistent nature of human rights enforcement fueled efforts to develop universal standards of justice. After World War II, the victorious states created international courts to deal with German and Japanese war criminals. The United Nations later established special tribunals to investigate and prosecute those responsible for atrocities committed in Rwanda, Sierra Leone, and Yugoslavia during the 1990s. Proponents heralded these tribunals as progress toward applying universal standards of justice, but these were relatively rare and ad hoc cases that relied on broad international consensus among UN Security Council members, especially the five veto-holding permanent members.

The establishment of the International Criminal Court (ICC) based in The Hague, Netherlands, was intended to address these deficiencies by creating a permanent mechanism for the indictment and trial of perpetrators of war crimes and other violations of human rights. Although the court's name suggests a broad purview and some advocated the court be granted universal jurisdiction, the ICC's actual power is rather limited. The court's jurisdiction covers only those states that ratify the treaty, but many, including China, India, Russia, and the United States, have declined to do so and thus remain outside the court's purview. Among states that have ratified the treaty, there is general agreement that acts of genocide, wars of aggression, and crimes against humanity fall under the jurisdiction of the ICC, but states commonly disagree on the specific definition of those terms. Governments remain reluctant to intervene in the internal affairs of other states even when faced with clear evidence of human rights abuses, such as in Xinjiang, China, since 2014, western Myanmar since 2017, Iraq and Syria since 2014, the Taliban's takeover of Afghanistan in 2021, and Russia's invasion of Ukraine in 2022.

The limited effectiveness of the ICC and human rights protections in general have led some to conclude that the territorial foundations of the modern state system are simply incompatible with the establishment of global human rights and therefore immoral. This argument is most developed regarding migration. Since states and NGOs are often prevented from crossing borders to address human rights abuses or even differences in quality-of-life issues, some have argued that developed countries have a moral obligation to relax migration controls and allow all law-biding migrants to enter, not just those classified as political refugees. This proposal would grant all persons the right to migrate to places offering higher standards of living and morally compel all governments to admit those migrants. The proposal is unlikely to be implemented, but the discussion highlights an increasing appreciation for the moral and ethical dimensions implicit in the varied processes of bordering.

NGOs are also increasingly active and effective cross-border players. NGOs play a significant role as advocates for press freedom, conflict resolution, economic development, environmental improvement, and women's rights. One of the most noteworthy NGOs is the International Campaign to Ban Landmines (ICBL), a coalition of hundreds of humanitarian NGOs cooperating to rid the world of landmines and cluster munitions. The coalition's efforts helped bring about the Ottawa Treaty banning certain landmines. More than 150 countries have now signed the treaty, although many leading powers have not, notably China, India, Russia, and the United States. In

10. De-mining work in Sri Lanka is part of a global movement organized by the International Campaign to Ban Landmines (ICBL).

recognition of its efforts, the ICBL received the Nobel Peace Prize in 1997. Several other NGOs have received Nobel Peace Prizes for their efforts to bring about cross-border change, including the World Food Program (2020), the International Campaign to Abolish Nuclear Weapons (2017), the National Dialogue Quartet (2015), and the Organization to Prohibit Chemical Weapons (2013). The role of NGOs is unquestionably significant, but it is imperative to recognize that they lack the resources and authority to respond effectively to the stark disparities in human rights and standards of living in the contemporary territorial state system.

Epilogue: A very bordered future

Borders are expressions of human territoriality possessing material and symbolic dimensions that influence people's daily lives and overall quality of life. Ranging from local to global scales, borders are understood as both formal and informal practices, institutions, and mentalities that differentiate between categories of places, peoples, and things. Rather than simple lines that demarcate place or divide space, borders are manifestations of power in a world marked by significant spatial disparities in wealth, rights, mobility, standards of living, and overall well-being. It is for these reasons that borders are such an important topic of study.

Given the variable nature of borders, attempts to predict future trends have obvious limitations. Yet after reviewing the historical evolution of borders and the spectrum of contemporary border research, it is worthwhile to ponder what the future might bring, however speculative that might be. The Westphalian notions of territorial sovereignty and rigid borders were never absolute in practice, but they have been, and in large part remain, the predominant organizing principles concerning the political division of world. Yet the international socioeconomic, political, environmental, and cultural linkages that characterize the processes of globalization seem to challenge this paradigm.

Some scholars and policymakers argue that globalization is creating a "flat world" of flows and networks that will gradually replace a bounded world of places. Would this lead to a more equitable and ethical territorial framework? Would it enhance overall human well-being through the more effective, productive, and sustainable distribution of labor, resources, and opportunities for education, healthcare, and expressions of identity? How would open borders/no borders impact indigenous sovereignties? How would environmental protections be implemented in the absence of bordered territory?

Given those uncertainties, some have worried that open borders or no borders would lead to widespread exploitation and a decline in social services because borders are so deeply entangled with the provision and regulation of labor markets, welfare, environmental protection, and political participation. Though far from ideal, state-based international norms have provided important foundations for responding to major human rights violations such as genocide and other atrocities, as well as providing humanitarian assistance after natural disasters.

The call for open borders or no borders is perhaps more prevalent now than in the recent past, but many scholars, pundits, and politicians discount those dramatic predictions and proposals, instead arguing that a bordered international system will remain the dominant modality of power, domestically and internationally, for the foreseeable future. From our perspective, reality is much more complicated than either the open-borders or walled-world views acknowledge. We see today, and are likely to continue to witness, contradictory trends regarding borders. Rather than remaining essentially static or being completely negated, current events suggest the territorial assumptions and role of borders in the contemporary world are experiencing a period of profound transition and renegotiation. There are clear historical precedents for such transformative periods. The emergence of new forms of socioeconomic organization fueled the initial formation of the territorial nation-state model in Western Europe, for example. Imposed through colonialism, that model gradually supplanted the relatively flexible notions of territory and borders that generally predominated in other regions.

It appears likely that we are experiencing a similar transition in the early twenty-first century as new modes of socioeconomic organization, activity, and identity emerge. Yet the result will almost certainly not lead to a complete de-territorialization and elimination of borders. Any de-territorialization will likely coincide with some type of broad-based re-territorialization. This will require corresponding processes of re-bordering, although they will differ in form, function, and scale from previous structures. Territoriality will remain central to human engagement with space, and as such the prospect of a continuous and permanent de-territorialization culminating in a borderless world is highly unlikely.

So, what might these new modes of territoriality and bordering look like? Given that borders are becoming more open to certain sets of people, institutions, and movements while simultaneously more closed to others, it is worth pondering if we are witnessing the emergence, or perhaps re-emergence, of more flexible, mobile, and unstable territorial networks like those that characterized medieval Europe and non-European areas before colonialism. This would imply modes of territoriality characterized by overlapping, contingent, and flexible hierarchies of political power. The growing size and influence of international corporations, supranational organizations, and nongovernmental organizations, as well as subnational groups like local governments, indigenous peoples, and diasporic communities, all challenge the notion of absolute state territorial sovereignty. Such trends, combined with state policies institutionalizing graduated sovereignty and citizenship and the proliferation of neoliberal economic spaces, could suggest the emergence of "neo-feudal" sociopolitical networks where certain classes and institutions enjoy broad privileges, while others face greater discrimination, regulation, and barriers.

During this stage of shifting spatiality, borders retain multifaceted and contradictory roles. Perhaps more prevalent than at any point in history, we are confronted with the challenge of reconciling the transportable and multi-scalar nature of territory, belonging, and governance with the reality of our very bordered world. In short, borders still matter and will continue to play powerful roles in global political, socioeconomic, cultural, and environmental issues. The task before us is to create "good borders" that protect those peoples, places, and things requiring protection, filter and block the nefarious, connect in positive and productive ways, and define and differentiate while also allowing the full realization of human potential. That we felt it necessary to put the term good borders in quotes suggests the complexity of the concept and the need to seriously consider varied perspectives on dividing space and creating place.

One clear means of doing so is through the rich literary and cinematic genres—spanning an array of novels, shorts stories, autobiographies, documentaries, short films, and full-length dramas—that explore the complex cultural beliefs, histories, and material circumstances affecting individuals and communities within border spaces. *Sicario*, *Traffic*, *La Niña*, *Babel*, and *Cartel Land* are just a few films about the Mexico/US border that are particularly useful in supplementing academic studies of undocumented border crossings, struggles for indigenous land rights, and deepening interdependencies among states, societies, economies, and ecosystems. These authors and filmmakers provide textured accounts of life on the border that highlight the realities and consequences of actions and inactions of policymakers, voters, and broader publics.

Borders have been, are today, and will remain central components of the human experience for the foreseeable future. The bounding of space is intrinsic to humanity. Humans are essentially place-makers, creating order by utilizing our capacity to physically, symbolically, and mentally demarcate differences between political, cultural, socioeconomic, and environmental processes, institutions, and networks. As a result, our world is crisscrossed by borders dividing varied spaces of authority, ownership, belonging, and opportunity. The field of border research provides a rich venue for understanding the changing nature of human social-spatial organization. It is imperative that we understand how borders are being reconsidered, renegotiated, and reformulated in contemporary socioeconomic, environmental, cultural, and geopolitical practices if we are to improve our individual and collective capacities for action amid the dynamics of globalization.

Further reading

Much of the material discussed in this book is drawn from research articles published in a variety of academic journals across the social sciences and humanities. Some of the best journals for border research are *Journal of Borderlands Studies*, *Geopolitics*, *International Studies Quarterly*, *Political Geography*, *Regional Studies*, *Journal of Ethnic and Migration Studies*, *Space and Polity*, and *Territory, Politics and Governance*. The list below, obviously far from comprehensive, focuses on some of the most important recent books in border research. Their bibliographies contain extensive citations to earlier works related to borders. Readers may also consult the Very Short Introductions on Diaspora, Empire, Geopolitics, Globalization, International Migration, International Relations, Nationalism, and Refugees.

Agnew, John. *Hidden Geopolitics: Governance in a Globalized World.* Lanham, MD: Rowman & Littlefield, 2022.

Anderson, Malcolm. *Frontiers: Territory and State Formation in the Modern World.* Cambridge: Polity, 1996.

Arts, Bas, Arnoud Lagendijk, and Henk van Houtum, eds. *The Disoriented State: Shifts in Governmentality, Territoriality and Governance.* Dordrecht, Netherlands: Springer, 2009.

Billié, Franck, ed. *Voluminous States: Sovereignty, Materiality, and the Territorial Imagination.* Durham, NC: Duke University Press, 2020.

Buchanan, Allen, and Margaret Moore, eds. *States, Nations, and Borders: The Ethics of Making Boundaries.* Cambridge: Cambridge University Press, 2003.

Cerny, Philip. *Rethinking World Politics: A Theory of Transnational Neopluralism.* Oxford: Oxford University Press, 2010.

Correa-Cabrera, Guadalupe, and Victor Konrad, eds. *North American Borders in Comparative Perspective.* Tucson: University of Arizona Press, 2020.

Dalby, Simon. *Anthropocene Geopolitics: Globalization, Security and Sustainability.* Ottawa: University of Ottawa Press, 2020.

Dear, Michael. *Border Witness: Re-imagining the US-Mexico Borderlands through Film.* Berkeley, University of California at Berkeley Press, 2023.

de Blij, Harm. *The Power of Place: Geography, Destiny, and Globalization's Rough Landscape.* Oxford: Oxford University Press, 2008.

Diener, Alexander C., and Joshua Hagen, eds. *Borderlines and Borderlands: Political Oddities at the Edge of the Nation State.* Lanham, MD: Rowman & Littlefield, 2010.

Diener, Alexander C., and Joshua Hagen, eds. *Invisible Borders in a Bordered World: Geographies of Power, Mobility, and Belonging.* London: Routledge, 2023.

Dodds, Klaus. *The New Border Wars: The Conflicts That Will Define Our Future.* London: Ebury Press, 2021.

Elden, Stuart. *Terror and Territory: The Spatial Extent of Sovereignty.* Minneapolis: University of Minnesota Press, 2009.

Espejo, Paulina Ochoa. *On Borders: Territories, Legitimacy, and the Rights of Place.* Oxford: Oxford University Press, 2020.

Friedman, Thomas. *The World Is Flat: A Brief History of the Twenty-First Century.* New York: Farrar, Straus, and Giroux, 2005.

Hastings, Donnan, and Thomas Wilson, eds. *Borderlands: Ethnographic Approaches to Security, Power, and Identity.* Lanham, MD: University Press of America, 2010.

Herb, Guntram H., and David H. Kaplan, eds. *Scaling Identities: Nationalism and Territoriality.* Lanham, MD: Rowman & Littlefield, 2018.

Houtum, Henk van, Olivier Kramsch, and Wolfgang Zierhofer, eds. *B/Ordering Space.* Aldershot, UK: Ashgate, 2005.

Jones, Reece, ed. *Open Borders: In Defense of Free Movement.* Athens: University of Georgia Press, 2019.

Jones, Reece. *White Borders: The History of Race and Immigration in the United States from Chinese Exclusion to the Border Wall.* New York: Beacon Press, 2021.

Jones, Reece, and Corey Johnson, eds. *Placing the Border in Everyday Life.* London: Routledge, 2014.

Kolers, Avery. *Land, Conflict, and Justice: A Political Theory of Territory.* Cambridge: Cambridge University Press, 2009.

Kütting, Gabriela. *The Global Political Economy of the Environment and Tourism.* Basingstoke, UK: Palgrave Macmillan, 2010.

Maier, Charles S. *Once Within Borders: Territories of Power, Wealth, and Belonging since 1500.* Cambridge, MA: Belknap Press, 2016.

Megoran, Nick. *Nationalism in Central Asia: A Biography of the Uzbekistan-Kyrgyzstan Boundary.* Pittsburgh, PA: University of Pittsburgh Press, 2017.

Mezzadra, Sandro, and Brett Neilson. *Border as Method, or, The Multiplication of Labor* Durham, NC: Duke University Press, 2013.

Morris, Ian, and Walter Scheidel, eds. *The Dynamics of Ancient Empires: State Power from Assyria to Byzantium.* Oxford: Oxford University Press, 2009.

Nail, Thomas. *Theory of the Border.* Oxford: Oxford University Press, 2016.

Newman, David, ed. *Boundaries, Territory, and Postmodernity.* London: Frank Cass, 1999.

Nicol, Heather, and Ian Townsend-Gault, eds. *Holding the Line: Borders in a Global World.* Vancouver: UBC Press, 2005.

Nikolic, Zoran. *The Atlas of Unusual Borders.* Glasgow: Harper Collins, 2019.

O'Leary, Brendan, Ian S. Lustick, and Thomas Callaghy, eds. *Right-Sizing the States: The Politics of Moving Borders.* Oxford: Oxford University Press, 2001.

Ong, Aihwa. *Neoliberalism as Exception: Mutations in Citizenship and Sovereignty.* Durham, NC: Duke University Press, 2006.

Paasi, Anssi, Eeva-Kaisa Prokkola, Jarkko Saarinen, and Kaj Zimmerbauer, eds. *Borderless Worlds for Whom? Ethics, Moralities, and Mobilities.* London: Routledge, 2019.

Phelps, James R., Jeffery Dailey, and Monica Koenigsberg. *Border Security*, 2nd ed. Durham, NC: Carolina Academic Press, 2018.

Popescu, Gabriel. *Bordering and Ordering the Twenty-First Century: Understanding Borders.* Lanham, MD: Rowman & Littlefield, 2012.

Rajaram, Prem Kumar, and Carl Grundy-Warr, eds. *Borderscapes: Hidden Geographies and Politics at Territory's Edge.* Minneapolis: University of Minnesota Press, 2007.

Sack, Robert. *Human Territoriality: Its Theory and History.* Cambridge: Cambridge University Press, 1986.

Sager, Alex. *Against Borders: Why the World Needs Free Movement of People.* Lanham, MD: Rowman & Littlefield, 2020.

Sassen, Saskia. *Territory, Authority, Rights: From Medieval to Global Assemblages*. Princeton, NJ: Princeton University Press, 2006.

Scott, James, ed. *A Research Agenda for Border Studies*. Cheltenham, UK: Edward Elgar, 2020.

Stein, Mark. *How the States Got Their Shapes*. New York: Smithsonian Books/Collins, 2008.

Summa, Renata. *Everyday Boundaries, Borders, and Post Conflict Societies*. Cham, Switzerland: Palgrave Macmillan, 2020.

Torpey, John. *The Invention of the Passport: Surveillance, Citizenship and the State*. Cambridge: Cambridge University Press, 1999.

Trigger, Bruce. *Understanding Early Civilizations: A Comparative Study*. Cambridge: Cambridge University Press, 2003.

Vaughan-Williams, Nick. *Border Politics: The Limits of Sovereign Power*. Edinburgh: University of Edinburgh Press, 2012.

Wastl-Walter, Doris, ed. *Ashgate Research Companion to Border Studies*. Farnham, UK: Ashgate, 2011.

Wilson, Thomas, and Hastings Donnan, eds. *A Companion to Border Studies*. Oxford: Wiley Blackwell, 2016.

Winichakul, Thongchai. *Siam Mapped: A History of the Geo-Body of a Nation*. Honolulu: University of Hawaii Press, 1994.

Yuval-Davis, Nira, Georgie Wemyss, and Kathryn Cassidy. *Bordering*. Cambridge: Polity, 2019.

Websites

Association for Borderlands Studies

www.absborderlands.org
Provides information about the association's publications, meetings, and other activities.

Border Criminologies

https://www.law.ox.ac.uk/centre-for-criminology/centre-criminology
https://blogs.law.ox.ac.uk/research-subject-groups/centre-criminology/ centreborder-criminologies/blog/2020/01/landscapes-border
A variety of resources focusing on border enforcement, including blogs and interactive map Landscapes of Border Control, sponsored by the Centre for Criminology at the University of Oxford.

Border Policy Research Institute, Western Washington University

https://wp.wwu.edu/bpri/
Supports research on a wide range of border issues with a focus on the Canadian/US border.

Borders and Boundaries by Joshua Hagen, Oxford Bibliographies

https://www.oxfordbibliographies.com/view/document/ obo-9780199874002/obo-9780199874002-0056.xml
An annotated bibliography surveying recent border research.

Borders in Globalization

https://biglobalization.org/
International partnership based in Canada promoting a variety of publications, events, and activities.

Centre for International Borders Research, Queen's University Belfast

https://www.qub.ac.uk/research-centres/bordersresearch/
Offers descriptions on the center's research, publications, and an extensive bibliography of border literature.

International Boundaries Research Unit at Durham University

www.dur.ac.uk/ibru
Includes current news database, conferences, and research focused on border conflict resolution.

Nijmegen Centre for Border Research, Radboud University

http://www.euborderscapes.eu/index.php?id=ncbr
Contains information on the center's research, conferences, and seminars.

UiT The Arctic, University of Norway, Border Poetics/Border Culture

https://en.uit.no/forskning/forskningsgrupper/gruppe?p_document_id =344750
Contains information on the center's research, resources, publications, and projects.

Index

Index

Index

CITIZENSHIP
A Very Short Introduction
Richard Bellamy

Interest in citizenship has never been higher. But what does it mean to be a citizen of a modern, complex community? Why is citizenship important? Can we create citizenship, and can we test for it? In this fascinating Very Short Introduction, Richard Bellamy explores the answers to these questions and more in a clear and accessible way. He approaches the subject from a political perspective, to address the complexities behind the major topical issues. Discussing the main models of citizenship, exploring how ideas of citizenship have changed through time from ancient Greece to the present, and examining notions of rights and democracy, he reveals the irreducibly political nature of citizenship today.

> 'Citizenship is a vast subject for a short introduction, but Richard Bellamy has risen to the challenge with aplomb.'
>
> Mark Garnett, TLS

GEOPOLITICS
A Very Short Introduction
Klaus Dodds

In certain places such as Iraq or Lebanon, moving a few
feet either side of a territorial boundary can be a matter of life
or death, dramatically highlighting the connections between
place and politics. For a country's location and size as well as
its sovereignty and resources all affect how the people that live
there understand and interact with the wider world. Using
wide-ranging examples, from historical maps to James Bond
films and the rhetoric of political leaders like Churchill and
George W. Bush, this Very Short Introduction shows why,
for a full understanding of contemporary global politics, it is
not just smart - it is essential - to be geopolitical.

'Engrossing study of a complex topic.'

Mick Herron, Geographical.

GLOBAL ECONOMIC HISTORY

A Very Short Introduction

Robert C. Allen

Why are some countries rich and others poor? In 1500, the income differences were small, but they have grown dramatically since Columbus reached America. Since then, the interplay between geography, globalization, technological change, and economic policy has determined the wealth and poverty of nations. The industrial revolution was Britain's path breaking response to the challenge of globalization. Western Europe and North America joined Britain to form a club of rich nations by pursuing four polices-creating a national market by abolishing internal tariffs and investing in transportation, erecting an external tariff to protect their fledgling industries from British competition, banks to stabilize the currency and mobilize domestic savings for investment, and mass education to prepare people for industrial work.

Together these countries pioneered new technologies that have made them ever richer. Before the Industrial Revolution, most of the world's manufacturing was done in Asia, but industries from Casablanca to Canton were destroyed by western competition in the nineteenth century, and Asia was transformed into 'underdeveloped countries' specializing in agriculture. The spread of economic development has been slow since modern technology was invented to fit the needs of rich countries and is ill adapted to the economic and geographical conditions of poor countries. A few countries - Japan, Soviet Russia, South Korea, Taiwan, and perhaps China - have, nonetheless, caught up with the West through creative responses to the technological challenge and with Big Push industrialization that has achieved rapid growth through investment coordination. Whether other countries can emulate the success of East Asia is a challenge for the future.

www.oup.com/vsi

INTERNATIONAL MIGRATION
A Very Short Introduction
Khalid Koser

Why has international migration become an issue of such intense public and political concern? How closely linked are migrants with terrorist organizations? What factors lie behind the dramatic increase in the number of women migrating? This *Very Short Introduction* examines the phenomenon of international human migration - both legal and illegal. Taking a global look at politics, economics, and globalization, the author presents the human side of topics such as asylum and refugees, human trafficking, migrant smuggling, development, and the international labour force.

INTERNATIONAL RELATIONS

A Very Short Introduction

Paul Wilkinson

Of undoubtable relevance today, in a post-9-11 world of growing political tension and unease, this *Very Short Introduction* covers the topics essential to an understanding of modern international relations. Paul Wilkinson explains the theories and the practice that underlies the subject, and investigates issues ranging from foreign policy, arms control, and terrorism, to the environment and world poverty. He examines the role of organizations such as the United Nations and the European Union, as well as the influence of ethnic and religious movements and terrorist groups which also play a role in shaping the way states and governments interact. This up-to-date book is required reading for those seeking a new perspective to help untangle and decipher international events.

www.oup.com/vsi